AN INVESTIGATIVE STUDY OF CONTRACT ADMINISTRATION PRACTICES OF GENERAL CONTRACTORS ON FEDERAL AND STATE DOT PROJECTS

———————————————

A dissertation

Presented to

The College of Graduate and Professional Studies

College of Technology

Indiana State University

Terre Haute, Indiana

———————————————

In Partial Fulfillment

of the Requirements for the Degree

Doctor of Philosophy

———————————————

by

George Okechukwu Okere

December 2012

Keywords: technology management, systems thinking, performance measurement, contract administration, change management, dispute resolution, claims resolution, partnering

ABSTRACT

Department of transportation (DOT) projects in the U.S. are plagued by issues resulting from poor contract administration performance. Literature reveals that there are unanswered questions related to contract administration practices and performance. Some of the most pronounced issues include construction disputes and litigation, failure rate of contractors, contractor misconduct and false claims, and the ability to staff projects properly. This study investigated the relationship between contract administration practices and contract administration performance of general contractors on federal and state DOT projects in the U.S. The overall research question addressed in this study was: "Is there a relationship between contract administration practices and contract administration performance of general contractors on federal and state DOT projects in the U.S.?" Data for this study was obtained from 20 state DOTs, and comprised of 86 samples.

Based on the research question, the study's hypotheses were derived from the literature, and a quantitative correlational research design method was used to investigate the relationship between the dependent variable (contract administration performance) and the independent variables (management attitude towards contract risks, contract provisions for mitigating contract risks, stability of scope definition, contract administration infrastructure, resource allocation strategy, and competency of contract administrators).

The first key finding was that a significant correlation existed between contract administration performance and resource allocation strategy. The second key finding from the study was that the average cycle time from discovery to execution of change order was two (2) months, and this can be used as the baseline for evaluating performance level. The third key finding from the study was that on average the practices in the questionnaire were applicable to more than 84 percent of the respondents, which confirms that the practices do apply to most state DOTs, and can be streamlined by each state DOT for performance evaluation.

The study's findings showed that there was no significant positive correlation between contract administration performance and management attitude towards contract risks, contract provisions for mitigating contract risks, stability of scope definition, contract administration infrastructure, and competency of contract administrators. A predictive model was not developed because an investigation using regression analysis revealed that the collected data were not suitable for development of a predictive model.

The collected data for this study shows patterns that support only one of the six hypothesized relationships and further study was recommended. Using power analysis, the sample size for this study was calculated to be 100 samples; however, only 66 of 86 collected samples met the requirements for use in inferential statistical analysis. It is expected that with a larger sample size, the variant scenarios and patterns will become evident, and a statistical analysis could confirm the relationships and a predictive model could be developed.

ACKNOWLEDGMENTS

It is a terrific opportunity, and a privilege to join the scholarly conversation in one's area of interest. This is an opportunity that few people are fortunate to have in their lifetime. Just like an artist uses a canvas to express his or her idea, the dissertation project gave me a means to design, and conduct a research investigation on a very important topic. This achievement is an accomplishment that will never be taken for granted and there is a lot to be thankful for.

The path that took me to this accomplishment is a winding path of life's journey with interrelated steps leading to this date, for which so many things had to be in the right place at the right time for this day to come, and even the various struggles and disappointments are blessings in disguise, because they also made it possible for this day to come.

I thank my parents for their value in education, and for providing the opportunity for my early education which has encouraged my quest for more knowledge. I would like to express my appreciation and thanks to my wife Thelma Okere for being very supportive and understanding as she watched me walk this journey.

I express my sincere gratitude to my dissertation committee members, Dr. Lee Ellingson, Dr. Musibau Shofoluwe, Dr. James Smallwood, Dr. Kathy Ginter, and Dr. Cindy Crowder, for their time and effort in helping me complete this dissertation. I extend a special thanks to Dr. Lee Ellingson, for his invaluable guidance. Special thanks to Dr. Kathy Ginter, for her guidance in statistical analysis, and for her time and effort in proofreading this work. I am grateful to my colleague, Dr. Andrea Ofori-Boadu, with whom I have collaborated on so many areas on this scholarly journey. Many thanks to all the state DOTs that participated in this research, and helped provide data for this research. A special thanks to the respective contact person at the

state DOT that helped coordinate and distribute the survey to the resident engineers.

TABLE OF CONTENTS

LIST OF TABLES

LIST OF FIGURES

CHAPTER 1

INTRODUCTION: CONTEXT AND DIRECTION OF STUDY

Performance improvement represents a critical and often difficult task of managing construction projects. Performance is about ability or rate of meeting specific purpose and objectives. For a project, the overall objectives include working within scope, meeting schedule, meeting budget, and performing quality work. One of the principles of performance improvement is that to improve a process, there should be an appropriate measure of the process, as a way to guide actions toward continuous improvement and consistency. The use of on time, and on budget remains a critical way to measure the performance of a project, but we must admit that the overall performance of a project is a result of the performance of various project components. In every key failure or success, there are small failures or successes contributing to it. Unlike project performance that can be measured directly by the ability to meet time and budget, contract administration performance as a phenomenon could not be measured directly, and neither is it feasible to measure contract administration performance by means of an experimental control group. However, the right metrics must result from a good understanding of the key practices or factors associated with contract administration performance, which in turn are associated with overall project performance.

The increasing value of expenditure by federal and state governments on construction projects requires efficient and effective management of contracts. Problems such as poor owner relations, poor change management, poor dispute and claims management still remain an issue in

2

today's contracting environment. A measure of performance is critical to organizational, project and business process. A measure of performance allows practitioners to know if objectives are being met, what areas need improvement, and how to remain successful. Within the context of state Department of Transportation (DOT), contract administration performance adds to the overall performance of a project, as well as the overall performance of the agencies.

Contract administration is a practice that is well known within the construction industry and especially with the state DOTs. Within these state agencies, contract administration encompasses the activities aimed at meeting compliance to contract terms and conditions, change management, owner relations, dispute resolution, and claims resolution at the project level. The ideal situation is one where general contractors develop and maintain the right processes, structure, and resources to manage compliance, changes, owner relations, disputes, and claims at the project level without allowing project issues to get out of hand. The ideal situation will reduce or eliminate expenditures associated with the use of alternative dispute resolution (ADR) or litigation. The ideal situation will also reduce the number of failing companies or those going out of business. Every organization recognizes the importance of delivering consistent performance results, yet the ability to fulfill this objective is sometimes undermined by the use of the wrong theory, tools, methods and techniques resulting from lack of knowledge on the systemic nature of contract administration. The cost impact of poor performance is evident, and it can be agreed that this is a problem that requires a solution.

BACKGROUND

What resources are available to general contractors who have problems achieving contract administration objectives consistently? Help is out there in the form of experienced and knowledgeable contract administrators, dispute and claims consultants, and contract administration manuals and guides developed by various state DOTs. Also, help is out there in the form of contract administration standards developed by several professional associations, contract administrator certifications, and literature on applied research projects conducted by industry experts on the subject of change management. According to Ren, Anumba, and Ugwu, (2000), with all the widely-written literature on the problem of construction claims, there is little proof of significant improvements on the issue of construction claims. Unfortunately, the industry still continues to spend billions of dollars on construction litigation and alternative dispute resolutions. A study conducted by Fullerton (2005) found that financial cost, time commitment, ability to pursue new jobs, mental anguish and damage to relationship were some of the negative consequences cited by most contractors that have used litigation as a way to resolve construction conflicts. Practices leading to poor contract administration performances are wasteful; they can wipe out profit margins; and they fail to sustain a project objective. Equally, the same practices will fail to sustain the construction industry and the economy because of their systemic and ripple effects. The U.S. construction industry plays a pivotal role in the health of the U.S. economy.

Economic Value of the U.S. Construction Industry

According to the U.S. Bureau of Economic Analysis report (2011) on value added by industry as a percentage of Gross Domestic Product (GDP), in 2010 the Construction Industry in the U.S. accounted for about 3.4 percent of the total GDP. GDP is the primary metrics used to measure the health of a country's economy. The GDP represents the total dollar value of all goods and services produced annually within the boundaries of each country. The value of construction put in place in the U.S. in 2009 as reported by the U.S. Census Bureau (2011) was $907,630,420,000, and the value in 2008 was $1,067,044,580,000 as reported in value of construction put in place. Let's look at some of the facts and figures directly or indirectly associated with the general contractors' contract administration performance.

Construction Company Failures in the U.S.

The contract administration performance of general contractors can be compared to trial and error and may be more about using the wrong tools for the job. Zazaian (2006) points out that the practice is that upon contract award, most parties focus on the day-to-day requirements of the project, and they lose sight of the terms and conditions that were agreed upon.

In 2008, the U.S. Department of Energy (DOE) released a report on challenges faced by the agency. Poor contract administration practices were on the list of the most serious challenges faced by the agency (DOE, 2008). Ability to staff the projects properly was identified as an area of concern. This type of challenge may lead to project failure, which may subsequently drive organizational failure. The National Association of Surety Bond Producers (2009) reported that of the 1,155,245 general contractors and operative builders, heavy construction contractors, and special trade contractors operating in 2006, only 919,848 were still in business in 2008, which results in a 20.37 percent failure rate. The report stated that, the key factors that led to the failure of these companies included unrealistic growth, performance issues, character issues, accounting issues, and management issues. Some of the root causes in relation to performance and management issues include inadequate training of staff members on company policy and operation, insufficient and incapable personnel at upper management or project level, and the ability to administer and collect change orders. Other issues include situations where one or more contracts have a claim, where the project was not completed on time, and where a company was continually involved in litigation.

Construction Litigation in the U.S.

In reference to a paper written by Michel (1998), the amount of money spent on litigation in the U.S. averaged about $60 billion every year, and nearly $5 billion of which was spent in the construction industry.

According to The National Academies (2007), the costs of resolving disputes and claims in the construction industry may total between $4 billion to $12 billion or more each year. Indirect costs associated with disputes and claims may include inefficiencies, delays, and lost opportunity costs of diverting productive employees away from profit-making activities to support litigation. Another indirect cost associated with disputes and claims is the cost of poor relationships between parties.

Contractor Misconduct in the U.S.

Poor contract administration performance is evident in the scenario where parties claim what they are not entitled to, or do not get paid for what they are entitled to, and someone gets short-paid. False claims cost the tax payer millions of dollars each year, and efforts by Project on Government Oversight (POGO) help to expose this unethical practice (POGO, n.d.). POGO provides such data aimed at exposing the practice, and increasing awareness on the consequences of false claims. According to the U.S. Government Accountability Office-GAO (2006), from 2001 to 2005, the U. S. Department of Defense (DOD) indicted 718 companies on contracting fraud. Some of the measures noted by the DOD for reducing contracting fraud, waste, and abuse included senior leadership, competent workforce, and surveillance of contracts from award to closeout.

Rippling Impact of Poor Contract Administration Performance on the U.S. Economy

The consequences of general contractors' poor contract administration performance are waste of resources, and time, on non-productive, and non-value added activities, and in effect work against sustaining the construction industry.

Construction companies are largely made up of specialty companies, most of which are small business companies. A 2009 report from Small Business Administration-SBA on "The Small Business Economy: A Report to the President," confirmed that the construction industry is dominated by small businesses, and more than 86 percent of the companies in the construction industry are considered small. In the report, SBA reported that in the U.S., in 2008 there were 627,200 estimated new businesses, 595,600 estimated business closures and 43,546 bankruptcies. As noted by the National Association of Surety Bond Producers (2009), the failure rate for the construction industry in 2008 was 20.37 percent. These failures translate to lost jobs and small business loan failures, both of which have a direct effect on the U.S. economy. The large failure rate means that most projects at some point may take on the risk of contracting with failing companies, which may eventually expose those projects to failure if not properly managed. The success or failure of the projects and the companies involved may partly be dependent on good contract administration. The failing construction companies create social, economic and psychological problems in the construction industry with rippling effects to other sectors of the economy.

STATEMENT OF THE PROBLEM

Consistently meeting contract administration objectives remains an elusive goal. Contract administration objectives are hardly ever met and even when some contractors meet them on one project, they lack the ability to sustain performance from project to project (Okere, 2010). A good percentage of construction companies lack the ability to meet project performance, resulting in some of them going out of business. The construction industry spends an enormous amount of money annually on litigation and alternative dispute resolutions. In relation to false claims, a company suffers serious consequence if it fails to conduct its business ethically. Unethical companies may face loss of customers, high employee turnover, low productivity, poor image, and may also face civil and criminal penalties.

This study centered around contract administration performance of general contractors on federal and state DOT projects in the U.S., and the problem of this study was the investigation of the general contractors' ability to meet contract administration objectives on federal and state DOT projects in the U.S.

PURPOSE AND OBJECTIVE OF STUDY

According to Socrates, "An unexamined life is not worth living." In light of the socioeconomic impact resulting from the phenomenon where general contractors lack the ability to meet contract administration objective consistently, the purpose of this study was to investigate and assess current contract administration practices of general contractors working on federal and state DOT projects in the U.S.

In line with the purpose of this study, the objective of this study was to determine if a relationship exists between contract administration practices of general contractors on federal and state DOT projects in the U.S., and general contractors' ability to meet contract administration objectives consistently. This research was developed around applied research in technology management, and it attempted to answer two key questions: 1) What factors are associated with the phenomenon where general contractors on federal and state DOT projects in the U.S. lack the ability to meet contract administration objectives consistently? 2) What best practices should be implemented so that general contractors on federal and state DOT projects in the U.S. will consistently meet contract administration objectives? Given the overriding objective of the research, the resulting findings of this study serve to:

1. Identify contract administration practices that are associated with general contractors' ability to meet contract administration objectives consistently on federal and state DOT projects in the U.S.

2. Examine if an association exists between the dependent and independent variables, and if so, to design a predictive model that can be used to predict and control general contractor's ability to meet contract administration objectives consistently on federal and state DOT projects in the U.S.

SIGNIFICANCE OF STUDY

The significance of this study relates to the impact or change in the way general contractors and owners on federal and state DOT projects will administer projects. The relationship between contract administration practices and contract administration performance affect overall project performance. Improved management of contract administration will result in improved performance such as:

1. Extensive work has been conducted regarding this subject, and this is evident in research papers published on subjects such as change order management, dispute resolution, partnering, and project performance. This study examined the issue of performance from a systems thinking view by showing factors that contribute to contract administration performance and how they relate to the overall performance of a project.

2. It is also expected that the principles that emerged from this study can be written into the contract specification, as a way for owners to improve and standardize their contract administration requirements. The ability to meet contract administration objectives consistently will be improved, by adding contract administration functional improvement languages to the contract specification, in an effort to meet needs, and solve problems in technology management and construction management.

3. Every business process has some inherent and unpredictable level of variations, yet most of the variations in a business process are predictable and can be associated with the practitioner's actions, the tools, and the method in use. This study provided a project control tool for practitioners to evaluate variations in contract administration processes, by making it possible to compare current performance with expected performance so as to determine whether the system is operating properly.

4. The study provides a tool for assessment of the level of maturity of an organization's current competency in contract administration. This assessment tool is based on the right technology, infrastructure and capability that provides for optimal performance and excellence based on the anatomy of contract administration. Maturity levels include Ad Hoc, Repeatable (basic), Defined (structured), Managed (integrated with other functional areas) and Optimized levels (Saxena, 2008).

5. The construction industry will be well served when general contractors and owners work to meet contract administration objectives consistently. This will help both parties control practices that are related to the contract administration performance, and the impact of such practices on the project.

6. Federal and state agencies, such as USACE (United State Army Corps of Engineers), California Department of Transportation (Caltrans), Texas Department of Transportation, and other federal and state departments of transportation could use the findings of this study to evaluate, monitor and control general contractors' contract administration performance. This is significant because it will help the agencies with specific metrics to evaluate contractors in the area of contract administration performance. This may also enhance the current composite metrics used by several federal agencies to evaluate contractors' performance at the project level (Appendix E) as a source for identification of contractors to be selected to provide a proposal on "best value" projects.

7. General contractors can apply the research findings in the form of project control tools. For example, a functional improvement checklist can be generated from the items (practices) that are a manifestation of the key variables. This checklist can be used by general contractors to outline the relationships between contract administration practices and objectives, and evaluate leading indicators of general contractors' ability to meet contract administration objectives consistently.

8. For the first time a complete and easy-to-use tool is available to practitioners for performance measurement of contract administration within the state DOT environment. This tool will enable practitioners to identify, evaluate, and control factors that may influence general contractors' ability to meet contract administration objectives on federal and state DOT projects in the U.S.

RESEARCH QUESTIONS

The following questions explained what this study aimed to learn and understand:

Q1: Whether and to what extent does management attitude towards contract risks relate to contract administration performance?

Q2: Whether and to what extent does contract provisions for mitigating contract risks relate to contract administration performance?

Q3: Whether and to what extent does the stability of scope definition relate to contract administration performance?

Q4: Whether and to what extent does the existence of contract administration infrastructure relate to contract administration performance?

Q5: Whether and to what extent does resource allocation strategy relate to contract administration performance?

Q6: Whether and to what extent does contract administrators' competency relate to contract administration performance?

HYPOTHESES

The study hypotheses were based on well-defined research questions and were simply stated to reflect one dependent variable to one independent variable. Also they were specifically stated without ambiguity on the variables and the population of interest. With respect to a population of interest consisting of general contractors working on federal and state DOT projects, the tentative theories of this study included the following**:**

P1 – Ha: Management attitude towards contract risks is positively correlated to contract administration performance

P2 – Ha: Contract provisions for mitigating contract risks are positively correlated to contract administration performance

P3 – Ha: Stability of scope definition is positively correlated to contract administration performance

P4 – Ha: Contract administration infrastructure is positively correlated to contract administration performance

P5 – Ha: Resource allocation strategy is positively correlated to contract administration performance

P6 – Ha: Contract administrators' competency is positively correlated to contract administration performance

A quantitative research method was used to collect sample data and statistically determine whether and to what extent a relationship existed in the population of interest.

DEFINITION OF TERMS

1. Performance Management

Performance management relates to controls in response to feedback or information on activities with respect to meeting customer expectations and objectives.

2. Contract Administration

Contract Administration is the process of managing all elements of a contract phases with the goal of meeting compliance, maintaining good owner relations, managing changes, resolving disputes, resolving claims and avoiding litigation.

3. Risk

Risk is defined as an uncertain event or condition that, if it occurs, has a positive or negative effect on a project's objectives.

4. Change

Change is defined as any event that results in a modification of the original scope, execution time, or cost of work.

5. Dispute

A condition where an issue or problem in which two or more involved parties disagree on its existence, which party was at fault, its impact, when it became an issue or what solution to take. A dispute is a disagreement between the contracting parties on a contract issue.

6. **Claims**

A demand or assertion by one of the contracting parties seeking, as a matter of right, the adjustment or interpretation of contract terms, payment of money, extension of time, or other relief with respect to the terms of a contract.

7. **Partnering**

Partnering is simply a way of conducting business in which two or more organizations make long-term commitments to achieve mutual goals. This requires changing traditional adversarial relationships into team-based relationships. Partnering promotes open communication among participants, trust, understanding, and teamwork.

ASSUMPTIONS

Assumptions are factors that, for planning purposes, are considered to be true, real, or certain without proof or determination.

1. The study assumed that the instrument developed to gather data for this study was consistent with the study's theoretical context. In other words, that the data collection tool and data analysis accurately captured core concepts in the study.

2. The study assumed that the respondents who completed the questionnaire were those that have experience in the concepts and principles of contract administration.

LIMITATIONS

Although this study was carefully prepared, several factors were beyond the researcher's control, and are considered as having the potential to affect the results of the study or how the results were interpreted.

1. The study was limited by the number of responses received as the response rate would affect the ability of the study to detect association and difference if they do exist.

2. Because of the limited time of this study, a cross-sectional instead of a longitudinal study was chosen. A study using at least four waves of measurement in a longitudinal design may provide a better distribution of sample patterns, and understanding of what is going on within the larger population.

3. The selection of sample projects that were included in the study was conducted by the state DOT representative, with the understanding that those projects were randomly selected.

DELIMITATIONS

1. The scope of the study was limited to capital improvement projects (major, non-recurring expenditures, such as roads and bridges) executed by the federal and state DOT project in the U.S.

2. The study was limited to projects that were one year or more into construction at the time of this study.

3. The study was limited to projects that had encountered and executed some change orders.

4. The scope of the study was limited to design-bid-build projects.

5. The study was limited to general contractors on federal and state DOT projects in the U.S. and the results of the study were only generalized to this group only.

CHAPTER 2

REVIEW OF LITERATURE: WHAT PRACTITIONERS THINK IS GOING ON

The objectives of this section are as follows: present "systems thinking" as a guiding and integrative theory for use in finding a solution to the problem of general contractors' ability to meet contract administration objectives consistently, discuss technology management, construction management, and project controls within systems thinking, and how they fit within contract administration. Also, this section presents the anatomy (function, structure, components, and principle of operation) of contract administration, discusses what federal and state DOTs view as objectives, and best practices of contract administration, and discusses theoretical and empirical literature on the relationship of contract administration practices, and corresponding objectives.

BUILDING UPON THE WORK OF OTHERS – IDENTIFYING GAPS IN KNOWLEDGE

This research builds upon the work of various research focused on construction performance. The work of Korde, Li, and Russell (2005) titled "State-of-the-art review of construction performance models and factors" extensively chronicled research in the last 20 years on construction performances and factors that are said to influence the performance measures. The authors looked at 122 research papers related to development of construction predictive models, and they found that there was no consensus on key measures and factors. The state-of-the-art review by the authors found that researchers aligned performance scope into three levels that include project level, group level, and activity level. It was also pointed out that most of the research on construction performance centered on the project level. Another key finding was that some of the research was designed as basic research instead of applied research, which makes them difficult for industry adoption. The lack of consensus on measures and factors stem from the fact that different researchers use different units of analysis; they define measures and factors in different ways; and the scope of research may not be clearly defined. An earlier work by Belassi and Tukel (1996) sought to address the problem of lack of consensus in identifying key success and failure factors. The authors developed a schema that aggregated factors within groups that include 1) factors related to project size and complexity, 2) factors related to competency of project manager and team members, 3) factors related to organization structure and support, and 4) factors related to the external environment. Chan, Scott, and Chan (2004), also sought to address the same issue of lack of consensus on critical success factors (CSF) and key performance indicators (KPI), and the authors developed a framework that grouped the factors into 1) project management actions, 2) project-related factors, 3) project procedures, 4) human-related factors, and 5) external environment. Korde, Li, and Russell (2005), by categorizing the performance measures into four dimensions that include productivity, time, cost, and overall performance, the authors found that the 122 research papers on construction performance generated 77 factors that were said to influence one or more of the four performance measures.

It is necessary to point out that a measure of contract administration performance was not identified in any of the 122 research work conducted over the last 20 years, and there is a gap in knowledge on contract administration practices and performance on state and federal DOT projects in the U.S.

SYSTEMS THINKING

Organizations and projects are systems that depend on and are depended on by other systems. The use of the right subsystem, environment, structure, processes, resources and interrelationships are required for a system to meet its objectives, and perform its function effectively.

"Systems thinking" is an interdisciplinary and multiperspective view of the environment. Systems thinking originated from systems theory, which originated from biological science, and was pioneered by Ludwig Von Bertalanffy. "Since the fundamental character of the living thing is its organization, the customary investigation of the single parts and processes cannot provide a complete explanation of the vital phenomena. This investigation gives us no information about the coordination of parts and processes" (Bertalanffy, 1972, p. 410).

Boulding (1956) explained that the goal of general systems theory is to provide a realistic perspective of the numerous relationships and complexities of the real world. As the world becomes more and more compartmentalized or specialized, people may lose perspective of the complex and intertwined nature of the world and these beliefs may limit the ability to solve problems. Systems theory takes into account levels of process relationships in any given environment, and systems theory provides a view for understanding an organization's business process and what is going on (Dawson, 2007).

In line with the systems thinking approach to managing an organization, Walker (2007) argued that organizational structure requires relationship between parties, interdependency of task and people, reduced degree of differentiation, and a high level of management integration. An ecosystem view indicates that everything in life is connected, and nature holds a network of relationships. This implies viewing life, organizations, projects and business processes in terms of connectedness, context and relationship. With systems thinking, it is quickly realized that parts of a system are better understood only within the context of the whole system, and that process in a functional unit or organization are networks of relationships embedded in larger networks (Capra, 2006).

Functions and Interrelationships of Systems

Processes do not exist in isolation, and their function depends on the interrelationship with other processes. Knowledge about how processes and their components relate to other processes may be required to predict the outcomes of various events and conditions within a given environment.

The constant dynamic of a system means that each relationship comes with risk, opportunities, impacts and consequences. However, if opportunities, impacts and consequences of each relationship can be forecasted, then better controls can be set to mitigate the outcome of those relationships. Organizations and projects exist in an interconnected and interdependent world, and what happens in one area may affect another area. Technology transfer, technology impacts, outsourcing, globalization, produce trading are all reminders of how interconnected the world is.

Organization as a System

A good understanding of construction context, process, and socio-economic environment is necessary for solving construction problems. Integration, consistency, collaboration, flexibility, standardization, timely and actionable information, and clarity of roles and responsibilities are just but a few attributes of a well-managed project. Why take a "systems thinking" approach to construction management? The construction industry continues to suffer from adversarial problems and continues to seek for a solution. When every player has a different objective or agenda and fails to work collaboratively towards a common goal, the effect is that minor issues and problems are not properly managed, but end up in costly claims and litigation. This happens when projects are not managed as an integrated network of various players coming together for a common goal with the ultimate result of providing value to all involved and sharing risk impartially. A project is a system made up of various functional units operating within a peculiar environment and require the right processes for converting inputs into outputs. Performance at the organizational or project level is a result of performances at various processes undertaken at the functional level. Project performance is a result of various functional area performances, and contract administration is one of those key areas. A good number of studies have been conducted by researchers to identify and predict construction performance (Korde, Li, & Russell, 2005). Performance can be evaluated at the project level, functional level or work activity level, and most performance related research in construction tends to center at the project level. While some researchers may choose to evaluate performance at the project level, it is important to show how those performance measures and factors that affect performance outcome derive from the functional level and work activity level of a project environment. Ling, Low, Wang, and Egbelakin (2008), evaluated project performance of international architectural, engineering and construction (AEC) firms in China based on project management practices adopted by the firms. The performance measures used by the authors include cost performance, time performance, quality performance, owner satisfaction and profit margin, and the factors the authors operationalized were mapped to various functional levels and work activity levels within a project environment.

Construction is about relationships and interactions. For example, a change in scope may affect various parties, processes and resources. Hence, construction operates in a dynamic environment. Construction management, requires a holistic practice, rather than reductionist practice, and entails the management of the interactions between all elements of a project. Construction involves many people, processes and knowledge, and successful projects are the ones that are consistently managed as a system, because construction is, in fact, a system. This statement is in line with Rummler and Brache (1995) on organizations as a system.

According to Rummler and Brache (1995), an organization is a processing system and will not be effective if it is not managed as a system, because just as any system, an organization behaves as a system regardless of whether it is managed as a system. In today's business world, conditions change daily, and change should be expected and planned for. In the construction industry it may be rare to find projects that are constructed exactly as they were planned because of the dynamic nature of construction projects. The importance and application of systems thinking in a construction environment may be critical to the successful performance of a construction project. General contractors need to be sufficiently adaptable to cope with a changing environment. Organizations are systems set up to achieve predetermined objectives through different functional units or departments. Organizational systems are made up of interrelated subsystems, processes, people, and knowledge. An organization that operates as a system may have a competitive advantage if it is adaptive and responsive to changing trends.

The Rummler and Brache (1995) model (Figure 1) is a depiction of an organization as a system that is based on the framework that an organization behaves as: 1) an adaptive processing system that, 2) converts various resource inputs, 3) into product and services, 4) which it provides to receiving systems or market, 5) that answers to shareholders, 6) and is guided by internal criteria and feedback, 7) but ultimately is driven by the feedback from its market, 8) and is influenced by the competition, from the market it draws on resources and in turn provides product and services to the market, 9) and operates within the influence of social, economic and political environment, 10) within the subsystems there are functions which exist to convert inputs into product and services, 11) and interprets and reacts to the internal and external feedbacks, so as to keep the balance within the internal and external environment.

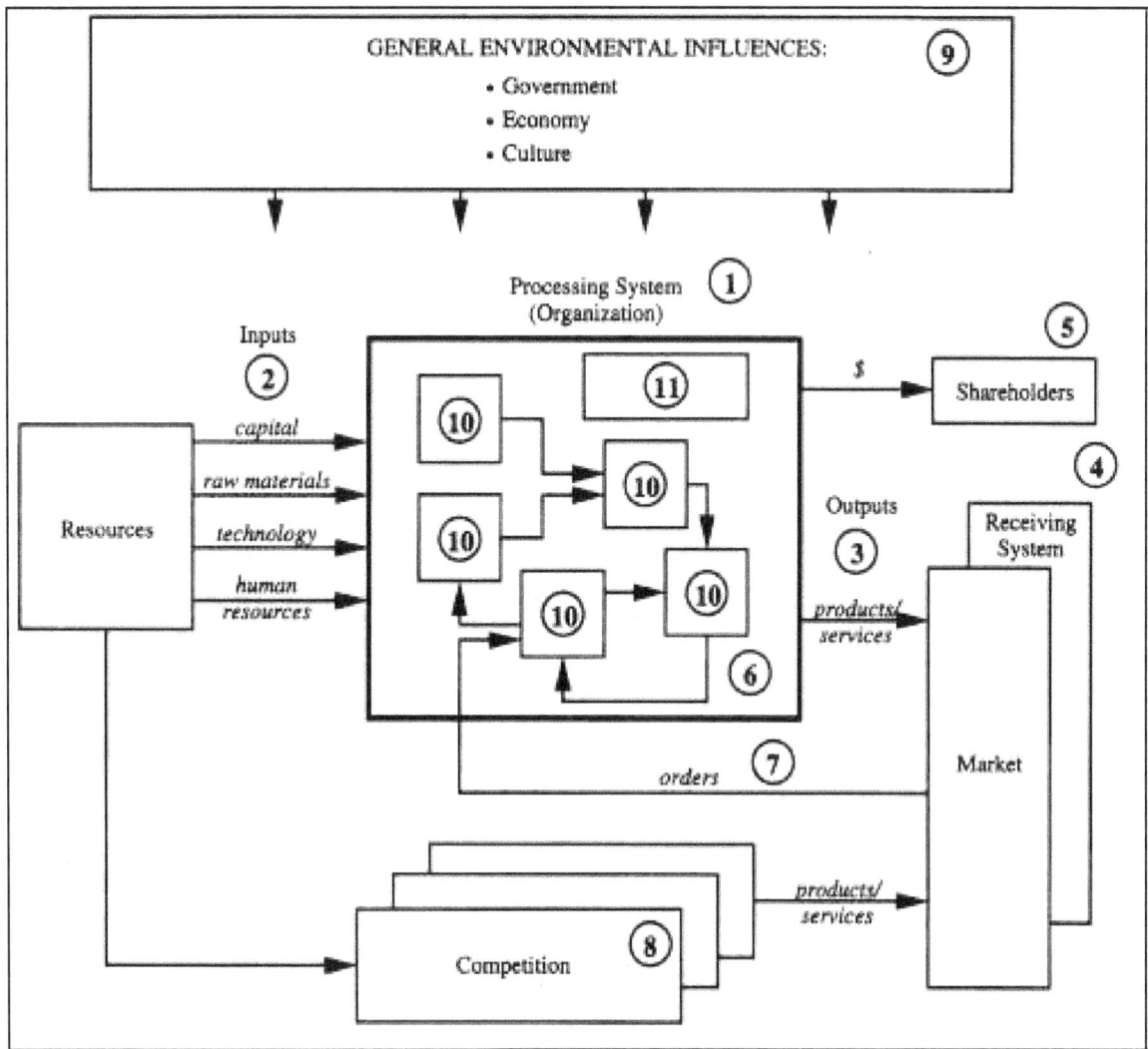

Figure 1. Rummler and Brache Model of An Organization as an Adaptive System

Whether it is about contracts for supply, service, manufacture, construction, research, or acquisition of "built environment," they all involve a common objective. The common objective is about assuring that the contract is administered to meet its objective without wasting money, resources and time. Contract administration is practiced within a broader context of technology management and construction management.

TECHNOLOGY MANAGEMENT, CONSTRUCTION MANAGEMENT AND SYSTEMS THINKING

Projects and organizations are systems and require the use of the right methodology and practice in order to manage these systems effectively. The practice of technology management and construction management contain the same fundamental principles that relate to "the universals of technology" (International Technology Education Association-ITEA, 1996).

Technology is ubiquitous in today's socio-economic environment and impacts everyone both negatively and positively. The idea that technology touches every aspect of life and is interrelated is an indication that it behaves like a system, and should be managed based on the systems thinking approach. Practitioners deal with technology every day in the form of tasks related to design, assessment, development, control, and utilization. What is technology and how can it be defined? Technology has many meanings within a variety of disciplines. To some it refers to a collection of systems; to others it opens possibilities; still to others technology means new tools and techniques. Some may believe that it opens the door to knowledge, while others see it as methods and processes by which transformation occurs in the form of results or outcomes that meet people's needs and wants. Technology as defined by Wyk (2002) is "Created competence. It is expressed in technological entities consisting of devices, procedures, and acquired human skills" (p. 19). It becomes necessary that practitioners have the skill and knowledge required to manage technology, and this is what management of technology (MOT) is all about (Khalil, 2000). Markert and Backer (2002) provided a perspective on technology with a view that technology is about human means developed in response to society's needs or desire to solve problems and in so doing may result in positive or negative impacts and consequences. A systems thinking view of technology is about impacts of technology and captures what Markert and Backer (2002) call characteristics of technology.

In an effort to meet needs and solve societal problems, technology has been developed, improved, and in some cases, created problems in the way humans inform and communicate, educate, travel, manufacture, and construct the built environment. Also, technology has been developed, improved, and in some cases created problems in the way society provides energy and power, explores space, fights wars, do business, practices religion, views cultural values, protects the environment, and cures disease. Every organization is involved in technological changes either as a creator, user or victim of technological innovations (Cardullo, 1996). This idea requires the need to understand and manage the effect of technologies and their dynamic environment.

Both goods and service industries undergo a series of production processes, which in fact involve relationships between various functions at several levels. The type and complexity of these relationships are such that they require practitioners to have a clear understanding of the system, process, knowledge, skill, and the tools and techniques that are used to create value, and maintain a competitive advantage. Production is the way goods and services are created, from the conceptual phase, through the implementation phase, to develop and market phase. At each phase, there are several parties involved and several agreements in place. How well the agreements are administered will determine the outcome of the production programs and projects. Contract administration as a functional element in managing technology is a key to an organization's future and its competitive advantage.

Every society needs knowledge and application of natural science, social science, applied science (technology), art, and humanities to solve problems and create value. Construction management is a tool that captures these areas of knowledge, and puts them to use to solve "built environment" problems, and create value to the entire society. Construction projects involve several potential players (Figure 2) that come together for one objective and one objective only, and that is to create value- value to the owner while providing for equitable compensation for all parties involved. Walker (2007) provided a definition of construction project management as:

"The planning, coordination and control of a project from conception to completion (including commissioning) on behalf of a client requiring the identification of the client's objectives in terms of the utility, function, quality, time and cost, and the establishment of relationships between resources, integrating, monitoring and controlling the contributors to the project and their output, and evaluating and selecting alternatives in pursuit of the client's satisfaction with the project outcome" (p. 5).

25

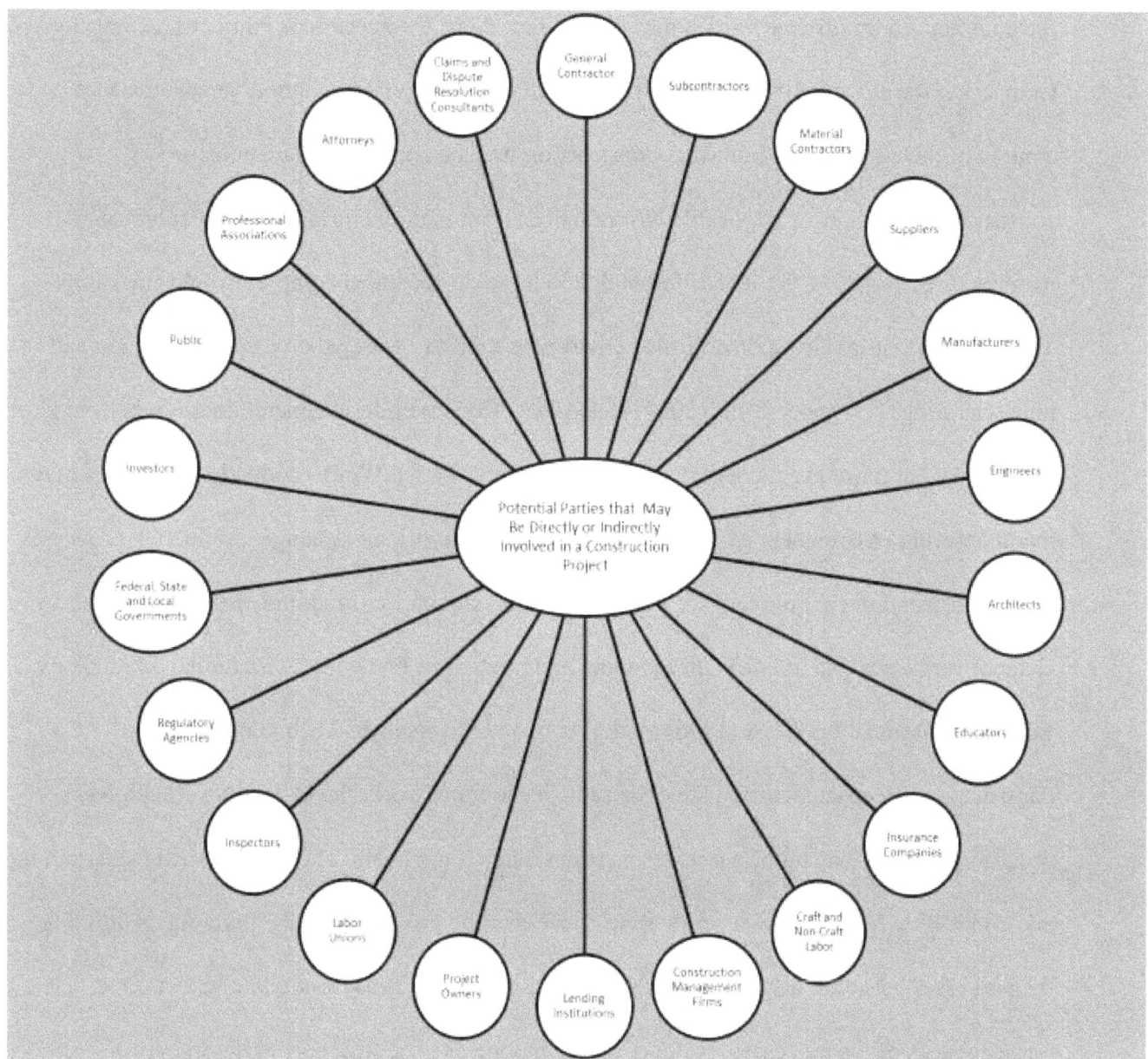

Figure 2. List of Potential Parties that May Be Directly or Indirectly Involved in a Construction

Project (created by Okere, O. G. to illustrate concept elements and the relationships of those

elements)

    While several functional areas are involved in making a project a success, contract

administration is a functional area that integrates with other areas, to meet project objective of

time, and budget. A project is a system, and the outcome of a project may depend on how well the contract is administered, which may in turn determine the success or failure of a company. Comparing the construction industry to manufacturing may provide a good perspective and context of the construction industry. Construction may be compared to manufacturing, construction may borrow applicable best practices from manufacturing, and incorporates manufactured products, yet it is different due to its socio-technical complexity and financing. The degree of variability and controlled environment differs widely between construction and manufacturing. Fernández-Solís (2009) stated that "Construction in general does not behave like an industry but more like a conglomerate of industries" (p. 3). While production processes in the manufacturing environment may encounter little to no variation or change, production processes in the construction environment are inherently full of stochastic (random behavior governed by chance) variations that are difficult to model and predict. In line with International Technology Education Association-ITEA (1996) model on "The Universals of Technology," both construction and manufacturing have the same basic framework, the first of which involves designing and developing of processes, systems, and components. The second one is determining and controlling how the processes, systems, and components behave. The third one is utilizing the processes, systems, and components. And the fourth one is assessing the impacts, and consequences of the processes, systems, and components. The fifth part of the framework involves knowledge of the nature and trend of processes, systems and components, and finally, the sixth part involves knowledge of the linkages between various processes, systems, and components.

To bring systems thinking into perspective within the context, this section illustrates a project scenario, and shows the network of parties that are involved every time a change is encountered.

Imagine that a change is encountered by a general contractor on a state DOT project in the U.S., and imagine that the change affects some subcontractors and vendors. Now, let's say that there is disagreement as to whether it is a change in scope, how to resolve it, which party is responsible for the change, and the cost of the change. Resources have to be redirected, assigned to manage and resolve the issue. Because of the associated costs, the stakeholders will have to be involved, and the lenders may be contacted to finance the additional scope. If the issue goes beyond a dispute, it will become a claim, and any of the party may choose to involve a dispute and claim resolution consultant. Through all of this, the contracting environment may shift from good owner relations and partnering environment to an adversarial environment as contracting parties may feel that they are not treated fairly. Also, the other thing that happens is that a contracting party, while looking to cut cost or stay afloat, may engage in poor workmanship, involve in false claims, or fail to meet contract terms and conditions. Depending on the conditions, a contracting party involved in this situation, may be forced to involve an attorney to start litigation, and may abandon the project, may require a surety company involvement. In other cases, a party may file for bankruptcy, or go out of business, and again the situation may require the involvement of an insurance company. Unfortunately, none of these options are desired as they have a negative effect on the contracting parties, their employees, the investors, the lenders, and a ripple effect on the city, state, and federal economy.

PROJECT CONTROL THEORIES AND MODELS: AN APPLICATION OF SYSTEMS THINKING

Systems thinking allows us to see relationships that are inherent in systems and between different systems, and provides a view of where we are and where we want to be. Understanding the gaps between current performance and ideal performance allows practitioners to see what, why and how general contractors meet contract administration objectives.

Within the context and application of systems thinking, project control theories and models are essential tools for use in evaluating and analyzing elements of a system that allows that system to meet or not to meet its objective. The need and application of systems thinking may be evident when evaluating changes and gaps in systems. Theories and models of project control allow the comparison of progress against plan so that corrective actions can be taken when deviations occur. Hanna, Camlic, Peterson, and Nordheim (2002) define change as "any event that results in a modification of the original scope, execution time, or cost of work" (p. 57).

According to Smith, Merna, and Jobling, (2006), the two most important fundamentals of control are that 1) control can only be exercised on future events and that 2) effective control allows for prediction of change. It follows that effective control turns risks into opportunities, and it all starts with a plan. According to Lewis (2000), without a plan, there is no control. Control involves monitoring in order to evaluate changes from the original plan, and thereby provide solutions to resolve the state of change.

A business process is made up of a series of steps (activities and tasks) that are completed to achieve process objectives in the form of products and services. In other words, a process transforms resources into results (International Standards Organization - ISO, 2001). According to Lewis (2000), work is controlled and not people. The author stated that control is all about comparing progress against the plan and making the necessary corrections when deviations occur. Lewis (2000) defined the control processes as involving a plan to detail the performance objectives, actual observation of performance, a comparison of planned performance versus actual performance, and the adjustment of the plan required to meet objectives.

Control methods provide for effective impact analysis. Impact analysis is a comparative evaluation of the plan as compared to the actual situation by determining the effect of changes on original business or on the operational plan. Change must be assessed by looking at business, or operational structures, processes, resources, procedures, tools and techniques, and conditions. Within the contract administration system, theories and models of systems thinking and project control provide a basis for understanding the elements, conditions and factors that go into contract administration. These items must be considered to understand the what, why and how general contractors meet contract administration objectives. It is about a systemic management approach as it relates to the absence or presence of the right structure, and the right processes for a contract administration system to meet its objective consistently.

According to Georgy, Chang, and Zhang (2005), project performance is dependent on the environment and conditions under which the project is conducted. In the case of contract administration, performance relates to meeting or exceeding objectives that include contract compliance, owner relations, change management, dispute resolution, and claims resolution. Practices that meet objectives consistently are sustainable practices, because they reduce and mitigate project risk by providing quality practices (get it right the first time), and by so doing, they reduce waste of time and money, and provide value. Whether they are public or private, large or small, organizations are set up to provide value to all stakeholders, and this is only possible when practitioners apply systems thinking, and incorporate the right processes (Rummler, Ramias, & Rummler, 2009).

A contract administration system that has the ability to meet contract administration objectives must have 1) foundational components, processes, tools, and techniques, and 2) efficient and effective application of these items in order to meet or exceed contract administration objectives. Consistency is about continuous improvement at all levels, and follows in line with Rummler and Brache (1995) view that successful performance improvement efforts must integrate three levels of performance that include organization, process and performer level.

ANATOMY OF CONTRACT ADMINISTRATION

Contract administration as a subsystem within an organization or a project is unique, and has a distinguishing characteristic that explains what contract administration does, how it does it, how well it does it, what it is composed of, and how it fits into the bigger system.

Following the discussions introduced above, what are the elements and attributes of contract administration? How is one technology differentiated from another? Each technology has distinguishing characteristics that are fundamental to that technology. To achieve simplification and a common understanding of the structure of technologies, Wyk (2002) provided a framework. Wyk's framework for defining characteristics of various technologies is referred to as "Strategic Technology Analysis (STA)," and involves (1) anatomy, (2) taxonomy, (3) evolution (life cycle), and (4) ecology of technology. Strategic technology analysis (STA) may be defined as an approach for evaluating technologies on the basis of their intrinsic characteristics (Wyk, 2002). The anatomy of contract administration is of interest here and, following Wyk's idea, this study identified the unique features that define the anatomy of contract administration by answering the questions in Table 1.

Table 1 Framework of Basic Features of Contract Administration (Adapted from Wyk's

Framework of Basic Features)

Characteristic	Question
1. Function	What does contract administration do? Refers to the role that contract administration plays within a given system
2. Principle of operation	How does it do it? Refers to the way in which contract administration performs its function
3. Performance	How well does it do it? Refers to how efficient contract administration executes its function
4. Structure	How is contract administration composed? Refers to the appearance and fundamental design of contract administration
5. Fit	What is the hierarchical position? Refers to the compatibility between contract administration and the greater system in which it is imbedded
6. Size	How large is contract administration? Refers to relative comparison of contract administration to other functional areas within the same project

Each of the above characteristics plays an important role in defining the fundamental aspects

of a technology, and the following section explains how they relate to the anatomy of contract

administration.

1. The function of contract administration within a project is to provide a methodology for

    adherence and compliance to contract terms and conditions, laws, regulations and

    performance. Adherence to contract is achieved through proper documentation, and

    execution of the contract in accordance with the contract requirements (scope, time,

    quality, and cost) as defined in the contract documents. Contract administration functions

    to make sure that all parties to a contract fulfill their obligations and earn their right to the contract. It is about making sure that both the owner and the general contractor have the relevant information to support entitlement and that the entitled party is fairly compensated for damages.

In executing the above function, contract administration as a functional unit within a project performs various operations and activities using the right knowledge and resources. The first stage is in understanding the contract, and what is contracted. Katz (n.d) stated that good risk management and the ability to meet objectives start with a good knowledge of the contract. Following Katz's recommendation, contract administration performs its function effectively by analyzing the contract to discover the troublesome clauses, and negotiate them as best as possible, and by responding to and watching out for the troublesome areas.

2. The principle of operation of contract administration involves several phases and processes (Figure 3). The process starts with contract execution, which is where an agreement is reached on what constitutes the contract scope, quality, budget and time, and contract requirements as shown in the contract documents. The next process establishes contract administration plans, identifies and assigns resources (people, knowledge, tools and techniques) required to monitor and manage all aspects of the contract. Perform contract is the third process, and this is where work is executed according to contract terms and conditions. In executing the contract, work must be monitored for compliance, changes, disputes, and claims, and this is the fourth process. The next process is the mitigation process, where prompt notice and collection of relevant documentation is crucial. Timely notification must be sent to allow the owner to plan and provide necessary directives and resources to manage changes. Also within this process,

proper documentation must be maintained to prove entitlement for damage. The sixth

process is the cost proposal, negotiation, and alternative dispute resolution (ADR)

process which require proof of entitlement and damage to validate claims for payment.

Next is the payment process, where claims for additional payments are made, and the

affected party is equitably compensated. Finally, the last process is the closeout process,

where all changes, disputes, claims are resolved, and both the physical and contractual

aspects of the contract agreement are fulfilled.

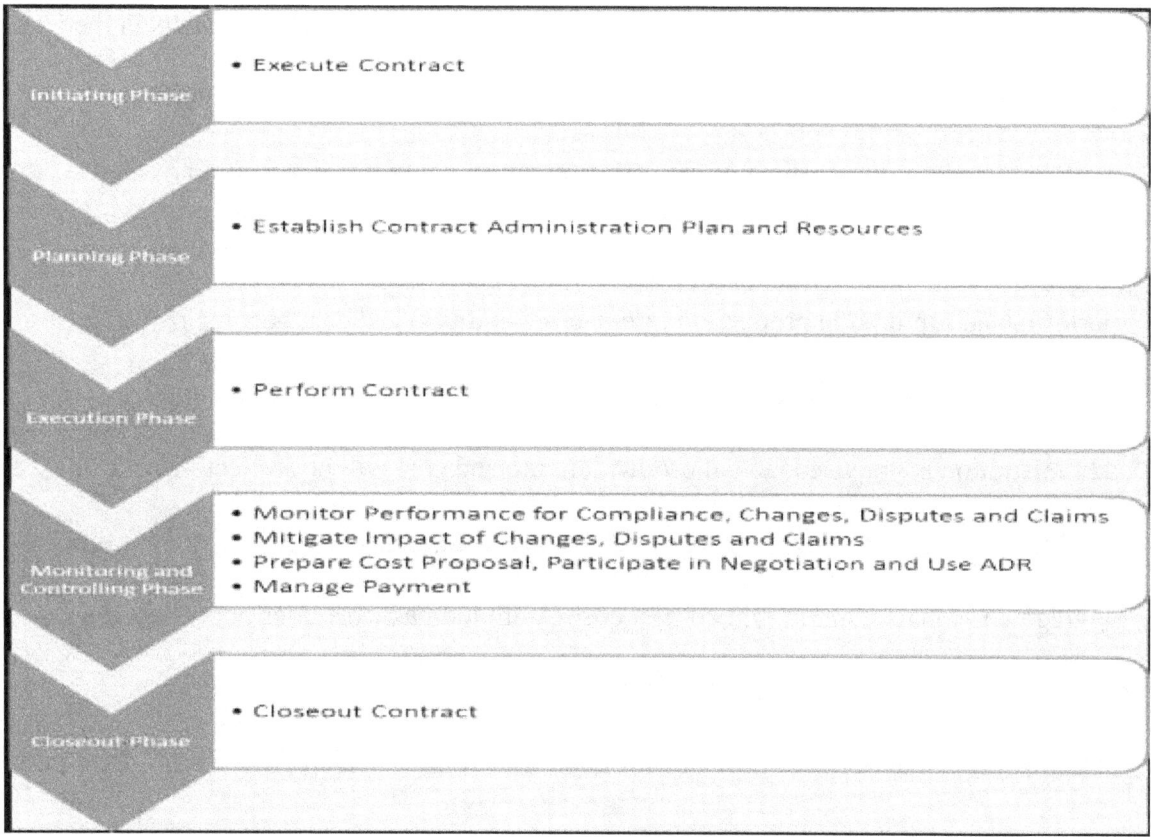

Figure 3. Processes Involved in Contract Administration Phases (created by Okere, O. G. to

illustrate concept elements and the relationships of those elements)

3. The goal of contract administration practices is to meet performance objectives and

provide consistent achievement of contract administration objectives. Effective contract administration allows for effective and consistent production, which allows for a successful project outcome. According to Park and Chapin (1992), a good cost estimate should provide for accurate projection of the cost of doing the work, and the best managed job should provide good past cost to help the contractor get new jobs. Lack of knowledge about the integrated nature of the construction system creates a problem and a hidden cost on projects and to the organizations. Informed decisions by managers should be based on the view of how each process integrates into the overall project environment.

Much of the general contractors' failure on federal and state DOT projects in the U.S. may be attributed to poor contract administration. An integrated systems view may explain this idea. Good estimates are estimates that are a realistic reflection of the expected conditions, and have fewer variances from actual conditions. Actual performance captured in production rate, factors, indices, and cost become part of past costs, which are then used as a baseline for future estimates. When poor contract administration is practiced, actual production rate and cost will not reflect the true condition. When scope changes are not properly managed and their costs are not properly aggregated, it may result in diluted past cost. A diluted past cost does not reflect the true condition, and is not a good baseline for future estimates. Effective contract administration management provides for good cost control, which provides for good past cost for use in bidding and wining new projects.

4. The various key elements that make up contract administration can be broken into fundamental structure which includes contract administration planning, performance evaluation, contract change management, disputes resolution, claims resolution, owner relations, payments, documentation, and closeout. Figure 4 provides a perspective of the

concepts, and components that relate to contract administration processes. Also, Figure 4

shows the interrelationship between all the components and subject areas in contract

administration.

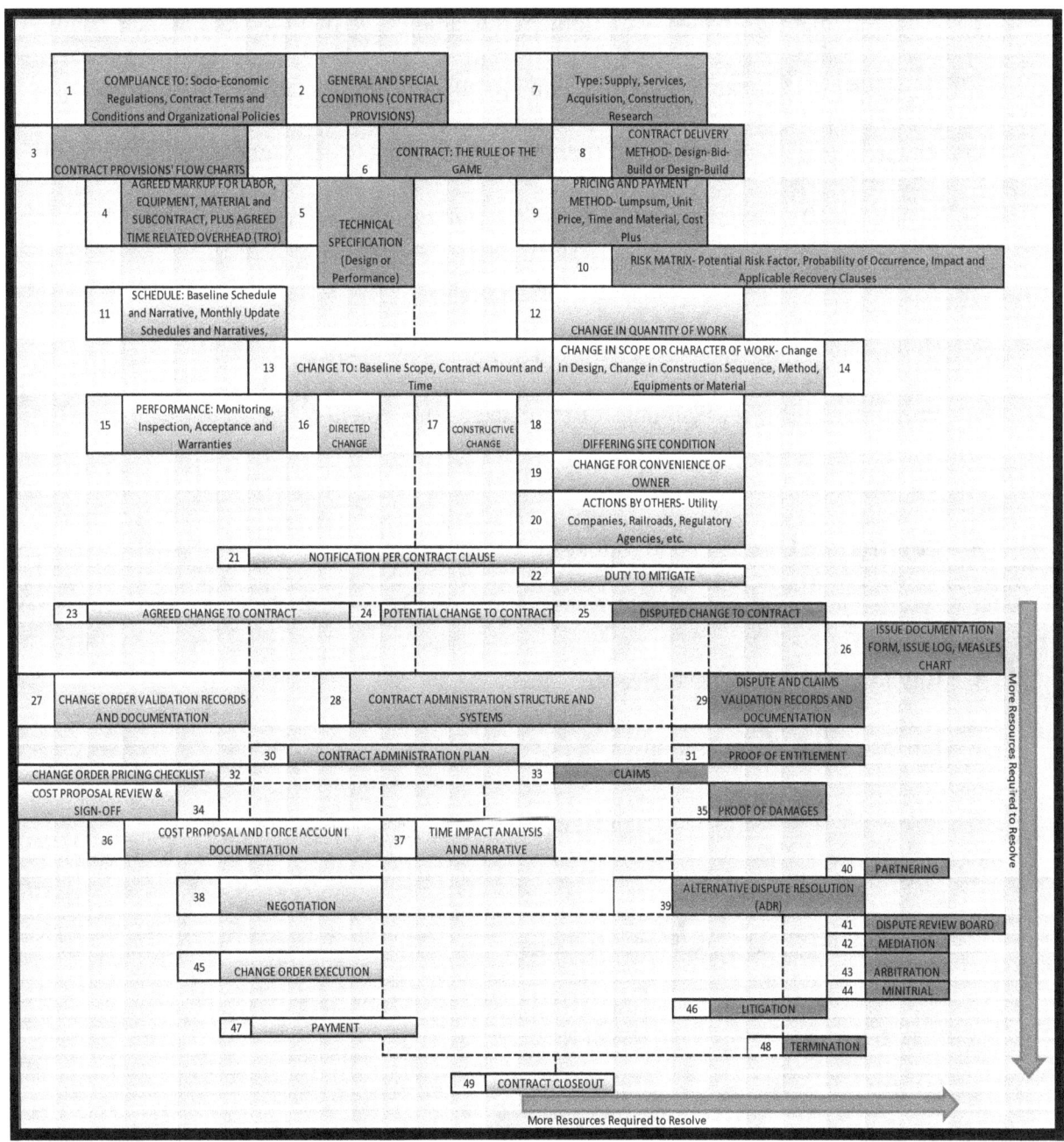

Figure 4. The Concepts and Components of Contract Administration Processes (created by

Okere, O. G. to illustrate concept elements and the relationships of those elements)

5. On any given project, various functional areas are organized to manage the project. Contract administration fits into one of the key functional areas within a project environment and constraints. In some companies, there are dedicated staff members on the project to manage change, while other companies have them located at their main office. Yet some companies have no team members dedicated to this function until they encounter change, which may lead them to engage claims consultants to help them out. U.S. DOE (2010), *Staffing Guide for Project Management*, identified the following Construction functional areas:

    - Contracting, Subcontracting, and Property Management

    - Program and Project Planning, Control and Management

    - Science, Engineering, and Design Support

    - Construction Oversight and Management

    - Quality Assurance

    - Environment, Safety, and Health

    - Finance and Administration

    - Safeguards and Security

    - Operations Oversight

    - Public Affairs and Stakeholder Relations

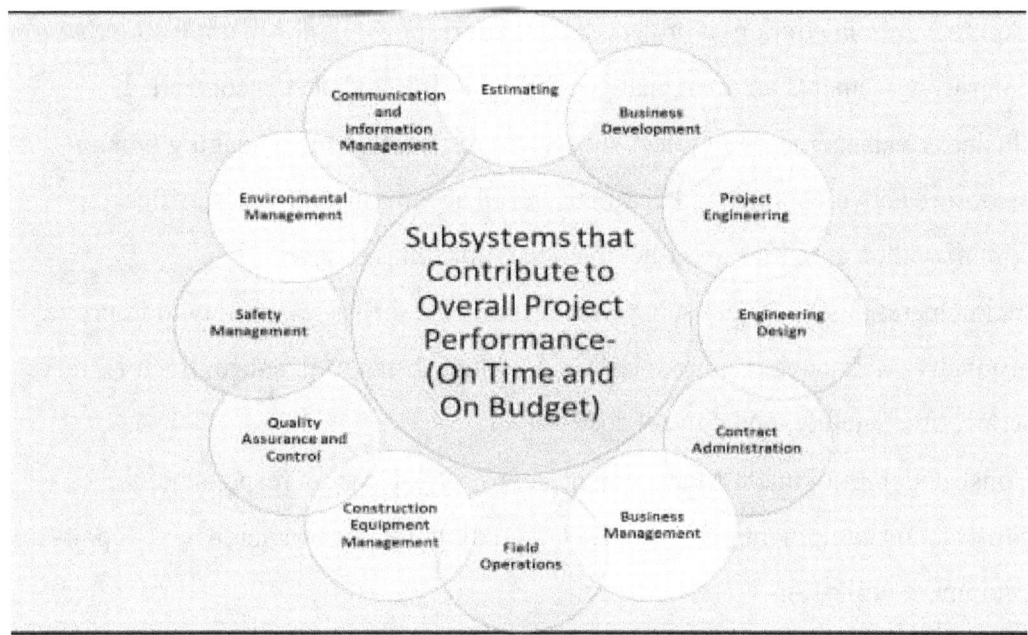

Figure 5. View of Project Performance and The Integrated Contributing Subsystems-Functional Areas (created by Okere, O. G. to illustrate concept elements and the relationships of those elements)

To put contract administration into perspective, some of the key functional areas of a project as depicted in Figure 5 include the following groups performing specific tasks and duties:

- Estimating- a project subsystem responsible for processes and activities for predicting the probable price of a project considering project dynamics, uncertainties, high degree of variability and unique conditions in location, time, environment, and resources.

- Business Development- a project subsystem responsible for tracking and identifying potential projects opportunities for bid and proposal.

- Project Engineering- a project subsystem responsible for setting up project, transforming the estimate price into a control budget, managing project documents, supporting field operations, and controlling scope, quality, time and cost.

- Engineering Design- a project subsystem responsible for research, conceptualization, feasibility study, development of design requirements, preparing preliminary design, preparing detailed design, construction planning, and construction oversight.

- Contract Administration- a project subsystem responsible for making sure that contract agreements are met, changes are managed, and cost is controlled.

- Business Management- a project subsystem responsible for managing human resources activities, financial and managerial accounting activities, office administration activities and tracking and controlling cost.

- Field Operations- a project subsystem responsible for issues on how to improve productivity, manage resources (labor, equipment, material, subcontractor), meet scope, time, quality, and control cost.

- Construction Equipment Management- a project subsystem responsible for construction equipment purchase, lease, rental, preventive maintenance, repairs and equipment utilization.

- Quality Assurance and Control- a project subsystem responsible for making sure that work is executed in accordance to contract specification, codes, and standards.

- Safety Management- a project subsystem responsible for creating a safe environment, making sure that people follow safe practices, and by so doing mitigate safety incidents, and control cost.

- Environmental Management- a project subsystem responsible for creating a safe, sustainable environment for all, and at the same time control cost.

- Communication and Information Management- a project subsystem responsible for acquisition, design, development, deployment and support of project information systems, and technologies.

According to Rad (2003), successful project performance is one where all the teams (functional areas) perform well at all phases, and within the context of changes to scope, schedule and cost. Equally important, is that, a good project performance evaluation tool, may provide an indication of what functions are critical to the overall performance of a project, and provide managers the basis to assign the necessary resources.

6. Contract administration function integrates to other functions and is not practiced in isolation. Each functional area relies on the others to get the job done, and the overall performance of a project depends on the performance of these functional areas. It is like a

symphony with sections and instruments: when they are in sync, everything works out well, but when one section is off, it distorts the entire sound. Each functional area is involved in the project as a producer and a user of project products, information, and services, and affects and is affected by the product or process of another. For example, past cost depends on actual cost at execution, which depends on proper tracking and aggregating of cost for original scope and changes to the scope. Contract administration provides a tool for effective management of various aspects of a project.

CONTRACT ADMINISTRATION OBJECTIVES OF FEDERAL AND STATE DOT PROJECTS

Contract administration is a fundamental part of a project, and has certain objectives. Within the state DOTs, what objectives should contract administration accomplish?

According to Garrett (2007) "Administering a contract entails creating a contract administration plan, and then monitoring performance through the many, varied activities that can occur during project execution. Key contract administration includes ensuring compliance with contract terms and conditions, practicing effective communication and control, managing contract changes, invoicing and payment, and resolving claims and disputes" (p. 37). From this perspective, contract administration can be defined as the process of managing all elements of a contract phases with the goal of meeting compliance, maintaining good owner relations, managing changes, resolving disputes, resolving claims and avoiding litigation.

From the perspective of state DOTs, what do contract administration objectives mean to them? Contract administration is a practice that is well known within the construction industry and especially with the state DOTs. Within these state agencies, contract administration encompasses the activities aimed at meeting compliance to contract terms and conditions, change management, good owner relations, dispute resolution, and claims resolution. According to the Office of Federal Procurement Policy (1994), the objective of contract administration is to create best practices that guarantee that the government receives goods and services on time and for the least cost, which means that taxpayers receive the best value. The purpose is to provide surveillance and mitigate risk so that products and services meet quality requirements, are on time, and within budget. The purpose of contract administration is to insure that a contractor is performing work in accordance with the contract.

The objective of contract administration is to provide technical oversight and direction as required. Technical oversight and direction includes confirmation that work has been or is being performed in accordance with the specifications and provisions of the contract, and includes provision of appropriate levels of monitoring, inspection, and acceptance as prescribed in the contract. Also, technical oversight and direction includes reviewing, approving, and monitoring payment for timeliness and accuracy, includes managing changes to the contract, and also includes documenting all actions taken with regard to the contract.

Several professional organizations (such as the Association for Advancement of Cost Engineering, and Construction Management Association of America) and government agencies (such as the USACE, and state Department of Transportation) have published standards that address the subject of contract administration practices, such as the "Contractor's Guide to Contract Administration" published by the USACE in 2002. According to State of Texas DOT (2007) "Contract administration is one of the most important jobs related to construction projects and involves numerous tasks occurring before and after contract execution and work order issuance. All work must be administered in accordance with the contract specifications, terms and conditions, state and federal laws and regulations, and department policy" (Chapter 1, p. 2).

The State of Louisiana DOT and Development (2011) adds that there are two parties to a contract and each party has rights and obligation. The contractor has the obligation to provide satisfactory performance, and in turn, the contractor has the right to fair treatment and prompt payment. A review of literature from some of the professional organizations and public agencies, such as the "Contractor's Guide to Contract Administration" by USACE indicates that effective contract administration embodies the area of contract compliance, change management, dispute management, and claims resolution, knowledge of the contract, change order payment, progress evaluation, owner relations and partnering.

CONTRACT ADMINISTRATION BEST PRACTICES PRESCRIBED BY FEDERAL AND STATE DOT AND AGENCIES

Contract administration is a fundamental part of a project and has some principles of operation and best practices to allow contract administration to perform its objectives. Within the state DOTs, what best practices are associated with contract administration?

An effective contract administration program is a risk management tool for both contractors and owners (Garrett, 2010). This section presents some of the best practices identified by various agencies, for providing some practical guidance and help to improve the contract administration performance. In 1994 the Office of Federal Procurement Policy published A Guide to Best Practices for Contract Administration, and Table 2 presents a summarized version of some of those practices.

Table 2 Summary of Best Practices Taken From A Guide to Best Practices For Contract (Source: Office of Federal Procurement Policy, 1994)

Areas of Concern	Summary of Listed Best Practices
Training of contracting officers	Use training and certification program to prepare for their roles and responsibilities.
Well-defined relationship	Establish partnership among practitioners to allow for synergistic outcome when team members work together.
Well-defined roles and responsibilities	Define project objectives, requirements and necessary roles and responsibilities to support project objectives.
Well-defined limitations of authority	Understand contract terms and conditions and know the scope, limitations and line of authority.
Adequate surveillance and monitoring of contracts	Develop contract administration plan with detail on how to plan, monitor and manage project changes.
Approval of contractor invoices and vouchers	Provide support for performance evaluation and ensure that payments are made to contractors on time, and in accordance to contract terms and conditions.
Management attention to contract closeout	Contract closeout starts as soon as the contract is executed and management involvement is required for planning and allocation of the right resources for physical item and contractual closeout.

Table 2 (Continued)

Areas of Concern	Summary of Listed Best Practices
Good management information systems	Proper documentation starts from the beginning and the right management information system must be used to manage all the items required for contract closeout.
Avoiding disputes in contract closeout	A contract is never closed out if there are disputes and claims waiting to be resolved. Partnering technique should be implemented to maintain a healthy contracting environment, and manage issues before they get out of hand.

The National Academies (2007), proceeding report presented some of the following best practices for reducing construction cost due to disputes and claims:

• Establish a cooperative project environment, with leadership from the top.

• Establish real time or job site techniques designed to get disputes resolved during construction.

• Assign project risk to the party that is best able to manage, control, and insure against the risk.

The U.S. Department of Energy (DOE) has implemented a series of significant contract and project management initiatives, including a root cause analysis (RCA) to identify the main challenges to planning and managing DOE projects. U.S. DOE (2008) report on Management Challenges at the Department of Energy indicated that one of the major challenges identified by the DOE is in the area of poor contract administration. In 2008, DOE wrote the "Root Cause

Analysis Contract and Project Management: Corrective Action Plan" and eight corrective

measures were identified and three of them related to:

1. Improving staff level through recruitment, training and retainage,

2. Strengthening risk management by providing risk management tools and processes, and

3. Improving oversight by providing structure and system for effective contract

   administration

The State of Texas Department of Transportation (2007) construction contract

administration manual covers both technical and administrative topics, and best practices. An

overview of the best practices for contract administration identified in the Texas Department of

Transportation (TXDOT) manual included:

• Complying with requirements stated in the contract documents and specifications.

• Enforcing state and federal regulations.

• Ensuring quality control by overseeing, inspecting and reviewing sampling and testing

of all materials and work.

• Keeping and maintaining accurate project records.

• Recording, verifying and preparing monthly pay estimates.

• Negotiating and processing of change orders.

• Promoting good public relations.

The TXDOT stated that the best practice is to maintain complete, clear and accurate

records that provide documentation of delays, quantity variations, work, quality of materials, and

other work records, to support contracting parties' position and right if and when disputes or

claims are encountered.

The State of Illinois Department of Transportation (2007) has developed a guide for best practices and created a construction inspection checklist for contract administration which included the following:

- Establish contract files and relevant documents.

- Receive a satisfactory progress schedule from the contractor prior to start of work.

- Make out a pay estimate at least once a month of the materials that are completed and in place, and the amount of work performed.

- Ensure that all materials incorporated into the work have evidence of the material inspection.

- Complete extra work forms on a daily basis, and establish agreement on unit price for the work.

- Ensure that the contractor is complying with all the equal employment opportunity (EEO) requirements.

The State of Minnesota Department of Transportation (2009) contract administration manual provided additional insight into some of the best practices of contract administration:

- Pre-construction conference with the contractor and all other interested parties as a way to validate that those requirements are followed.

- Protection of public interest in public buildings and other public works projects by making sure that the contractors are complying with labor related regulations.

- That the contractors make provisions that allow for clear and logical audit trail on changes that occur on the project.

The State of Minnesota Department of Transportation best practices also stated that diaries should be kept to provide a complete narrative picture of the project. Keeping project records is considered to be one of the most important duties and responsibilities of a contract administrator.

The U.S Army Corps of Engineers (2002) contractor's guide to contract administration provides a framework for contractors to meet their rights and obligations. Some of the best practices identified in this manual include:

- Use pre-construction conference.

- Provision for documentation of contract processes.

- Provision for proper control and management of subcontractor.

- Compliance with labor laws and submittal of certified payrolls.

- Development and maintenance of project progress schedule.

- Proper documentation to accompany payment requests.

- Proper process to be followed for contract changes based on the applicable contract clause or clauses.

The preceding report on contract administration best practices taken from various agencies captured many areas of interest to both general contractors and owners on practices that could help general contractors meet contract administration objectives, or in other words, how best to practice contract administration on federal and state DOT projects in the U.S.

The best practices identified above can be categorized into six main topical areas that include:

1. Management involvement,

2. Well-defined contract terms and mitigation strategies,

3. Stability of scope definition and requirements,

4. Infrastructure for tracking and documenting the contract process,

5. Resource allocation, and

6. Knowledge and competency of administrators

RELATIONSHIPS BETWEEN CONTRACT ADMINISTRATION PRACTICES AND OBJECTIVES

The preceding sections provided different perspectives on systems thinking, project control theory, anatomy of contract administration, contract administration objectives, and best practices. This section is centered on empirical and theoretical evidence to further support and explains why and how contract administration practices relate to contract administration objectives or performances.

Poor performance can be a result of various scenarios. A study by Visser and Chermark (2009) found that scenarios are useful in explaining complex business cases, environments, conditions, trends and interdependent factors that underpin the scenarios. The performance of a business process could be improved through detailed evaluation of business scenarios that constrain the business process. Figure 6 shows the six key scenarios discussed in this section that include management attitude towards contract risks, contract provisions for mitigating contract risks, stability of scope definition, contract administration infrastructure, resource allocation strategy, and competency of contract administrators.

In this section, the relationships and the effects of various contract administration practices on the ability to meet contract administration objectives were evaluated. An in-depth literature review was conducted; however, the review found that the available literature on contract administration does not fully address most of the relationships and effects resulting from the practices identified here.

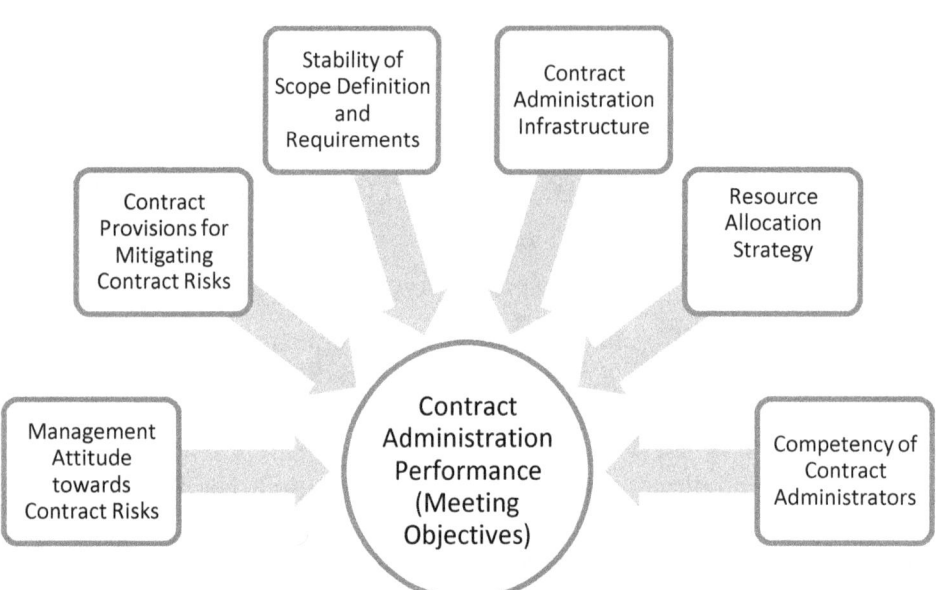

Figure 6. Relationship Between Contract Administration Performance and Contract

Administration Practices (created by Okere, O. G. to illustrate concept elements and the

relationships of those elements)

Management Attitude towards Contract Risks

A study by Hassanein and Afify (2007) on international contractors' perception and

attitude towards construction risk, as reflected by the number of exceptions taken in a bid, found

that risk identification by contractors differs by nationality. Having the right perception about

project conditions and risks should set the stage to implement the right tools for mitigation or

improvement of project outcome while there is still time. According to Mehta (2008), a project

control system with the main function to monitor and control project objectives is deemed a

failure if the system fails to aid in meeting overall project objectives of on time, and on budget.

Perception, belief, and behavior, go hand in hand. For example, general contractors that fail to

perceive and appreciate the consequences of poor change management are most likely to be

unprepared. This type of behavior may indicate failure to view projects in a holistic way, and

could be an indication that they lack systems thinking approach to contract administration.

Construction projects involve risks, and only when the risks are accurately identified can they be priced out accurately (Hassanein & Afify, 2007.)

The use and practice of ADR as a risk mitigation tool for resolving disputes and claims may depend on management ideology, belief, risk perception and attitude towards construction contract administration. Goetz and Gibson (2009) evaluated the trend of construction litigation on three government agencies that include General Service Administration (GSA), U.S. Naval Facilities Engineering Command (NAVFAC) and the U.S. Army Corps of Engineers (USACE). The authors found a significant reduction in the number of construction litigation from the projects. The authors attributed the reduction in litigation to increased use of partnering, design-build, and ADR techniques by these organizations. In resolving disputes and creating a mutual environment through partnering, it is necessary that the numbers of partnering meeting sessions be increased during heightened strain in the relationship between owner and general contractor. Also, periodic team building activities are encouraged among owner and contractor employees (Cheng & Li, 2002). The guide by the State of California Department of Transportation - Caltrans (2008) on partnering encourages team building activities, such as barbecue, bocce ball games, and family picnics. Cheung and Suen (2002) emphasized that strategic dispute resolution practices are very important, as unresolved disputes on a project will result in project delays, increased tensions, and poor owner relations. According to Nael (2003), partnering provides a strategy on how to stop a simple problem from getting out of hand.

It has been found that in the construction industry, the trend indicates a reduction in construction litigation (Goetz & Gibson, 2009). Though the cost is less if disputes are resolved using alternative dispute resolution methods such as arbitration and mediation, it is cost effective

50

to resolve disputes at the project management level without allowing it to escalate into the use of ADR or litigation.

Positive management attitude towards the use and practice of partnering is undoubtedly one of the most effective and inexpensive methods of managing construction risk. Yeung, Chan, and Chan (2007) used the Delphi method to identify key metrics to measure the successful implementation of partnering on a project, and the metrics included (1) time performance, (2) cost performance, (3) top management commitment, (4) trust and respect, (5) quality performance, (6) effective communications, and (7) innovation and improvement. Top management commitment remains at the top of this list because, without management support and involvement, the other items become difficult to attain. It is expected that the above criteria would help managers monitor and manage their partnering performance. The work by Yeung, Chan, and Chan (2007) also pointed out that quality, claims occurrence and magnitude, dispute occurrence and magnitude, litigation occurrence and magnitude, and other measures have been used in other literature to evaluate the success of partnering projects in construction.

Policies guide organizational objectives and must be used to define management's visions and how to achieve those visions. Policies are organizational governance statements aimed at meeting regulatory and organizational objectives and must be clearly stated. Cardullo (1996) stated that "policies are rules or guidelines that express the limits within which actions should occur" (p. 53). Well written policies also have detailed processes for compliance as well as a reference to the regulations or objectives that inform the policy. A written contract administration policy provides a guide for contract administrators to know the purpose of an effective contract administration and what processes must be implemented. As part of an organizational policy, the need to establish and maintain a contract administration department may indicate risk perception. A contract administration department may be established to respond and manage compliance and changes to avoid things from getting out of hand. This practice is expected of all parties involved in a project and not limited to the owner, the general contractor or the subcontractors. For example, a subcontractor's risk perception and risk management attitude may be evident in their ability to manage project issues. In other words, ability to administer contracts should be a key performance indicator.

A proactive risk management plan identifies contract clauses with potential risk and corresponding risk management steps to mitigate the impact on the project before they occur. How risk is perceived may define a belief system, which may define how risk is managed. Knowing that every project may encounter some change (variation from plan), allows managers to prepare and provide the right resources. For example, if a company's data indicate that revenue grows 5 percent to 15 percent on average on most projects, the company's ability to provide contract administration resources right from the start of each project may indicate risk perception of that company. Cox (1997) stated that most projects will not achieve successful management test if they were measured based on being free of change, because few projects will pass the test. A good start for an effective contract administration is to accept the fact that most projects will experience a change in scope. This perception must be founded on the belief that a project is dynamic and progressive just like the environment under which it exists. There is no perfect plan, no perfect owner, no perfect contractor, and no perfect condition.

Construction is a risky business, and there are many tools available to reduce the risk of failure, and poor performance, yet the problem persists. Shofoluwe and Bogale's (2010) study of risk management practices of major U.S. construction companies found that 95 percent of those surveyed have a risk management practice in place. Some of the risk items identified in the study included defective design, incompetence of subcontractors, claims and disputes, and differing site conditions. Differing site conditions relate to issues where the encountered physical condition is materially different from the conditions thought to exist at the time of bidding, and could not have been discovered by a reasonable site investigation. With most major U.S. contractors practicing risk management, why is it that they fail to meet contract administration objectives? Realistic risk perception and mitigation strategies provide for an effective contract administration practice as a tool to manage and reduce construction risk. Adams (2008) argued that most construction risk assessments are subjective often because of lack of good historical data, and this situation accounts for why contractors use contingency as a "catch-all" for risks. A study of risk identification practices in the United Kingdom found that contractors predominantly use subjective methods and checklist when identifying risks, as opposed to rigorous and systematic approach (Adams, 2008). The study by Adams supported the view that contractors still rely on subjective methods when evaluating construction risks. In other words, even though most U.S. contractors practice risk management, they may be using the wrong tool and methodology to manage risk, which may affect contract administration performance.

In response to the high risk associated with construction disputes and claims, El-adaway and Kandil (2009) explored a risk retention approach with the use of dispute and risk insurance provided by a risk retention group (RRG). It was expected that this approach would be beneficial to contracting parties in relieving the financial burden associated with direct and indirect effects of managing and resolving disputes and claims. The main idea being that after the contractor is paid by RRG for the claim, the RRG will then be left to pursue the claim with the owner.

Contract Provisions for Mitigating Contract Risks

Sanders (2004) argued that even though most practitioners accept the view that change happens, the practice indicates that there is no alignment in the acceptance of this view and how practitioners prepare for change. Sanders stated that a look at the contract indicates if there is true preparation on what to do under various contract change conditions or issues. Sanders pointed out that most contracts fail to indicate the contractual requirements that state what needs to be done when change happens, and this result in poor change management and resolution of issues. For example, a project could encounter a differing site condition, have defective and deficient contract documents, be suspended, encounter a labor strike, encounter a delay in delivery of owner furnished equipment, and experience adverse weather. Also, a key player could file for bankruptcy, the contract could be terminated, the key player may have superior knowledge and fail to share with others, and/or administer contract poorly. Any one of the above change conditions could fall under a directed change or constructive change or cardinal change. When the contract does not spell out what needs to be done when a change condition is encountered, time and resources are wasted while waiting to reach an agreement on what needs to be done. A good review of the contract allows for the general contractor to identify missing requirements and address the issues with the owner for earlier review and resolution before changes are encountered. Effective risk management is about being proactive instead of being reactive, because a proactive approach allows for monitoring and mitigation of risks (Kaliprasad, 2006c). A study by Ibbs and Ashley (1987), demonstrated that the way a contract clause is written may have either a positive or negative impact on the overall project performance. However, a well-conceived contract strategy relies on understanding the direction of impact of the contract clauses, and taking the right steps to administer the clauses and mitigate their impacts.

Contracting parties use exculpatory clauses to protect themselves, mitigate contractual risks and in turn expose or transfer those risks to others (Fisk & Reynolds, 2010). Most contracts still retain some exculpatory language that disclaim liability for differing site conditions, geotechnical study reports, responsibility to verify all dimensions and conditions prior to bid. However, as soon as the contract is signed, all contracting parties are expected to follow the terms of the contract while using the right practices to take advantage of non-exculpatory clauses, and take steps to negotiate and mitigate the effects of the exculpatory clauses. According to Katz, (n.d), a prudent contractor is one that understands the risk associated with each contract provision and takes the right steps to negotiate, transfer and manage the risks during performance.

In taking advantage of the non-exculpatory clauses while meeting contract administration objectives, a general contractor should always use agreed labor and equipment rate and markup (overhead and profit). The general contractor should reach an agreement with the owner on time related overhead (TRO) and what should be included when computing the daily time related overhead rate. A Caltrans (2006) memorandum on this subject differentiates what is cost-related overhead and what is time-related overhead. TRO is used to cover costs for time extension and can range from one (1) percent to 22 percent of the contract cost (Caltrans, 2006). Also, contracting parties should nominate and get approval for dispute resolution board (DRB) members so as to timely convene them to review and address disputes and claims. The nature of a construction project is such that changes are bound to happen, but how the changes are managed will make a difference. Levin (1998) argued that partnering and DRB support non-contentious project relations, which allows contracting parties to manage changes without being adversarial.

State DOT projects may have partnering provisions, and contracting parties are required to comply with the requirements. Some of the key elements of partnering include synergy, a common set of goals, and ability to solve problems collectively. Glagola and Sheedy (2002) found that partnering participants sought to achieve good relationships, dispute resolutions, and claims avoidance when they entered into a partnering workshop. Use of partnering encourages cooperative approach to project management, and by so doing, reduces poor relationships that create barriers between contracting parties. Communication barriers between contracting parties reduce common goal attitudes towards dispute and claims resolution. The us versus them attitude is strongly discouraged, and the contracting environment should be such that it allows for solving problems as one team with one agreed upon objective.

Owners' oversight and evaluations may be seen as intrusive; however, knowing the owners' hot buttons, expectations and level of satisfaction are a means to help isolate problems before they get out of hand. Owners' oversight and evaluation should be encouraged, welcomed and implemented at all stages and phases of any project. Periodic evaluation from the owners helps to evaluate how well a general contractor is meeting compliance. For example, state DOT projects with federal aid are required to add form FHWA-1273 (required contract provision federal-aid construction contracts) into subcontract provisions instead of making references to the form (Caltrans, 1997). Owner review of subcontract agreements executed by the contractor will allow the owner to assess this federal-aid compliance.

No general contractor plans to fail, but fail they do, and there must be an explanation for this phenomenon. Do they know the right intervention needed to resolve each problem? Failure is not an option that any general contractor would choose. Zazaian (2006) maintained that strategic contract administration practices start in the early phase of contract drafting, to ensure that the critical provisions relating to contract compliance, notifications, and project monitoring are clearly defined, and set forth in the finalized contractual document.

General contractors should establish policies on a false claims act in relation to cost proposal preparation, cost proposal submittal, and request for payment. Responsible corporate citizens are those that have integrity, understand the consequences of unethical conduct, and take every step to avoid unethical behavior. The contract language has been the basis to shift risk, from one party to the other. This also has resulted in the adversarial nature of the construction industry, as parties are left to tackle the uncertainties involved in the construction process. This problem is made worse due to stiff competition where some general contractors may underbid just to win a contract with the hope to make up through change order. Vanden Bosche (1981) stated that the three main causes of claims in the construction industry include contract communication failure, absence of integrity, and greed by the need to survive.

In meeting contract requirements, a general contractor should prepare a cost proposal based on an agreed and acceptable pricing method supported with accurate documentation, and should not prepare a cost proposal using estimated cost and estimated production rate when actual cost and production rates are known. Estimates contain assumptions, uncertainties and a high degree of variability based on a project context as to resources, location and time, and this explains why the value arrived at in an estimate is not a deterministic value but instead a probabilistic value (Dysert, 2007). Negotiation time could be reduced when price is based on current cost data instead of bid estimate data, because the level of accuracy in a bid estimate may not truly reflect the actual project conditions.

Contract administrators working on state DOT projects should be knowledgeable of and in compliance with regulatory requirements such as "Buy American" requirements. Requirements like these are set by laws and must be followed. Compliance includes both internal and external components. External components involve government laws and regulations, while internal components involve contracts, policies, and best practices. Most of the compliance items identified in the contract reference back to government regulations for which non-compliance is not an option. According to Criss (2006), compliance relates to quality control, audit, corporate governance, ethics and common sense. Criss (2006) pointed out that compliance requires a good understanding of client requirements, the contracting environment and constraints.

In meeting contract administration requirements, general contractors must meet contractual requirements, which include both internal and external components. A good indicator of general contractors' ability to meet requirements could be found by evaluating if a general contractor has no more than one open and unresolved notification from the owner on non-compliance to certified payroll, and if a general contractor has sponsored a subcontractor's claim without proper review.

Stability of Scope Definition and Requirements

Change management is a key part of risk management, and it involves reporting, controlling, documenting, pricing, and negotiating cost of changes (Gray & Larson, 2006). Gray and Larson pointed out that most changes can fall under three categories: 1) scope addition changes, 2) contingency changes implemented because of risk events such as design error and design omission, and 3) value engineering. The way these changes are managed may differ, but what is common is that they all require a well-defined scope.

Definitive planning is possible when scope and requirements are well defined by the owner, and scope is stable to the extent that the requirements are not changing as the plan is implemented. Many project disputes arise from design error and changes. A study by Hassanein and Nemr (2009) found that most change order claims result from concurrent design and construction process. Construction disputes can be reduced or eliminated by working to stabilize requirements and allow for proper planning and cost estimate before actual work starts. Delano (1998) indicated that the stability of requirements remains one of the key factors that contribute to project success, and the author stated, "Just as it is difficult to hit a moving target, it is difficult to manage a program that lacks stability" (p. 44). Project scopes need to be completely defined before work starts because every time a change is made to the scope, applicable resources must be changed. In other words, a well-defined scope allows for effective management of time, cost and quality (Khan, 2006). Scope requirements should be fully defined before attempting to come up with the strategies and cost estimate needed to accomplish the requirements. A comparison of a design specification and a performance specification might be used to drive down the point about scope definition. The level of risk associated with a design specification is less than the level of risk associated with a performance specification. As the level of scope definition increases, so does the degree of accuracy in determining the range of outcome.

The loss of a team member puts a company at risk because operations may be affected until a competent replacement comes up to speed (Schleifer, 1990). A general contractor with a lower contract administration team turnover (longer average tenure with the team), is expected to have less disruptive processes from team changes, and is also expected to have cohesive communication, and knowledge transfer. Investment in human capital is the key to organizational success, and it is about attracting the right people and retaining them through motivation, education and training (Kaliprasad, 2006a).

Essens, et al. (2005), identified team composition over time as an important determinant of team effectiveness. The authors argued that the longer the average time team members are together as a team, the higher their chance of being effective in meeting their objective. Since the day-to-day work of team members is interdependent, it means that their performance influences and affects each other, and the entrance or exit of a member may affect knowledge transfer. This illustrates that high employee turnover rate can be disruptive and may not allow for proper transfer of both tacit and implicit knowledge.

Meeting contract administration objectives are possible when requirements are well defined, clear and stable with fewer and less complex or disruptive variations. Owners need to have a dedicated design and engineering team to analyze, and define problems promptly before construction starts. Owners should also avoid situations where requirements change significantly over time while construction is in progress.

Contract Administration Infrastructure

A well designed infrastructure forms the foundation and framework for management of contract changes. According to Okere (2010), effective transfer and maintenance of any technology depends on the right infrastructure. As shown in Figure 7, the foundation of reliable performance starts from the presence of the right infrastructure.

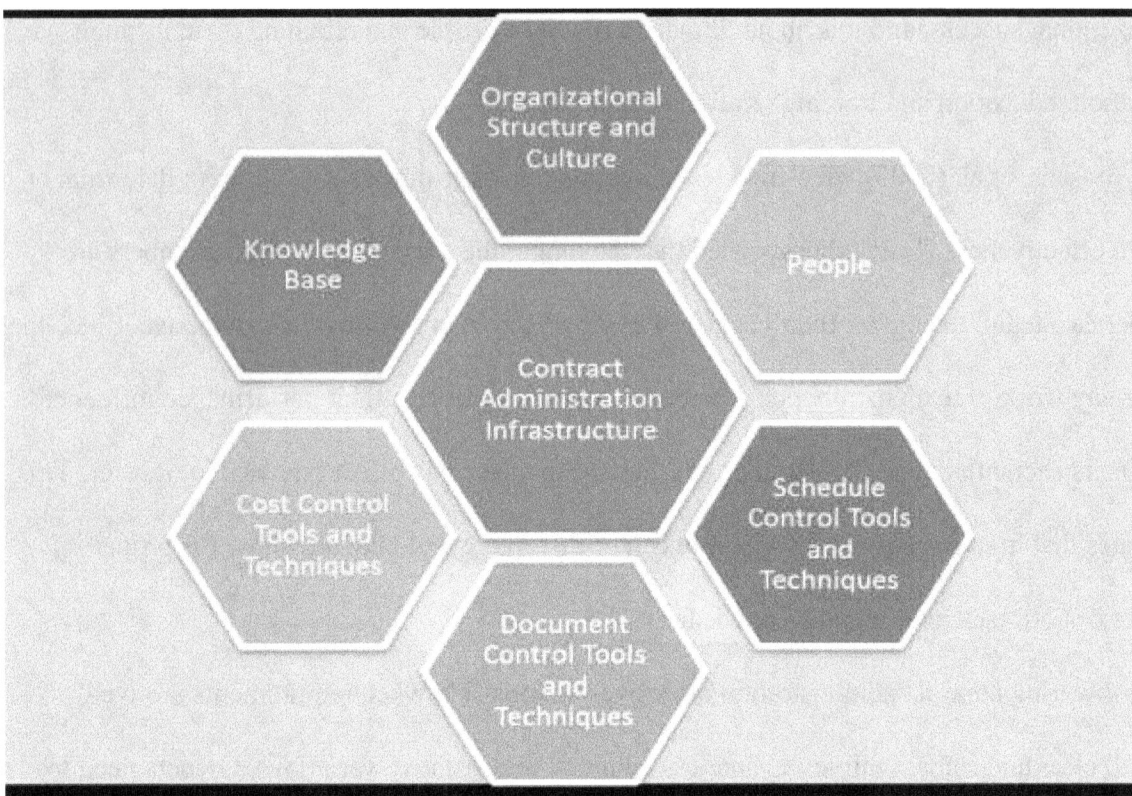

Figure 7. Elements of Contract Administration Infrastructure (created by Okere, O. G. to illustrate concept elements and the relationships of those elements)

The use of a personal computer in the construction industry has the advantage of helping to manage volumes of information created and used on a project. Today's construction companies may need to maintain an electronic document management system (EDMS) for effective documentation of construction processes and activities, including changes when they occur. Proper documentation of the construction processes is a major element of a successful project. Bauer (2005) pointed out that recovery for claimed damages is only successful when they are proven to a reasonable certainty. In other words, claims must be supported by documented evidence of the facts. Nalewaik (2011) argued that while good testimony is important in a claims situation, accurate documentation remains the strongest basis to prove entitlement. Regarding lessons learned, Schieg (2007) argued that post-mortem analysis (PMA) serves for the collection of experiences in organizations, and allows for identification and processing of project experiences gained and lessons learned during the construction of a project. Some of the advantages of PMA include identification of structures for process improvement process, and planning of improvement measures (Schieg, 2007).

In addition to documenting the activities required to execute original contract items of work, changes do happen and must be documented for proper proof of damage and entitlement. Charoenngam, Coquinco, and Hadikusumo (2003) proposed the use of a web-based change order management system (COMS) for proper documentation of contract changes. The conventional paper-based method is time consuming and inefficient for today's fast paced and complex project. Also, the use of an electronic ad hoc change order documentation method is no better than the paper-based method. A well-organized EDMS is the solution, and proper documentation of change order related activities cannot be overemphasized. Proper documentation remains one of the most important actions required to prove and validate change. Some of the documents that must be maintained to prove or validate a change include 1) contract agreement, 2) original contract documents, 3) RFIs, 4) time cards, 5) notices of change, 6) daily diaries, 7) executed change orders, 8) revised contract documents, 9) original estimate, 10) site investigation reports, 11) as-built, 12) notice of potential claims, 13) cost and financial reports, 14) equipment utilization schedule, 15) meeting minutes, 16) procurement records, 17) field production records, 18) project baseline and update schedules, 19) schedule narratives, 20) cost proposals, 21) photos and videos, 22) issue documentation log, 23) potential change order log, 24) change order log, and 25) progress payments. Documenting the construction processes and activities as situations unfold is more important than having a claims consultant and an attorney on your side. Schleifer (1990) pointed out that the industry has become very complex and has made way for services provided by construction attorneys, claims consultants and dispute avoidance specialists. However, without the right document, a claims consultant or attorney will lack the tools to prove entitlement and damage.

An equally important system that general contractors are required to maintain is a cost control system. The cost control system is used to establish a budget and track costs against the budget for performance measurement and changes. A cost control system provides the tool for managerial and financial accounting. A poorly structured cost control system may fail to isolate changes when they occur, and this could lead to arguments on actual cost and validity of what is included in cost for changes. A good cost control system is one that allows for forecasting of cost and completion time of the project, based on detailed construction schedule, and quantity-based control budget for progress measurement (Floyd, 2004). An effective cost control system should be such that allows for reporting on operation production rates, and should be capable of aggregating cost per original scope, agreed changes, and potential changes. According to Norfleet (2005), productivity is a measure of performance as a function of quantity of work produced per labor effort spent. Ability to manage production can only be achieved when variances are identified on time, allowing for corrective measures to be taken (Tichacek, 2006). However, the right cost structure must be set up to track variances. Cost structures should be set up for equipment and labor utilization at the lowest task level, and this makes for isolation of cost at the task level. Another functional element of a cost control system is that control budgets are well detailed to allow for variance identification of out-of-character work and cost. A well-developed control budget provides for effective and efficient analysis and identification of changes, as well as providing proper data for validation of entitlement. According to Tichacek (2006), a three step methodology for cost management must involve measures and strategies aimed at managing 1) accurate and complete baseline estimate, 2) production and productivity, and 3) change and scope.

Cash flow is about getting paid for work completed, but most importantly it is about getting paid on time. An effective contract administration system is one that controls those items that could delay timely payment, and thereby provide the right structure to avoid late payments from happening. A study by Rwelamila, Lobelo, and Kupakuwana (2004) found that inabilities of organizations to meet their financial obligations are in most cases due to inadequate cash resources. One area that general contractors may fail to pay attention to is the area of proper documentation of a time and materials contract. Unlike lump sum and unit price contracts, time and material (T&M) contract requires documentation of every labor, equipment, material used and the corresponding authorization and signature from the owner. An effective contract administration system will be one that is designed to make sure that T&M sheets are signed on a daily basis. Timely payments make the difference between negative and positive cash flow. Force account calls for proper documentation of equipment and labor rates and their utilization (Fisk & Reynolds, 2010). A general contractor should be no more than one (1) month behind in billing and receiving payment for actual work completed on executed change orders. Equally important is that the owner is prompt at review, approval and payment of a progress request without incurring interest for late payment.

Resource Allocation Strategy

The need to assign knowledgeable field personnel to supervise, track, document and validate change order work cannot be over emphasized. Human resources remain the most important asset an organization could use to strategize its objectives and competitively position itself for the future. However, certain basic practices must be in place if a company must tap into their human resource capabilities. Availability and timely assignments of the right resources are the key to effective managing of projects. Huselid (1995) found that such practices as recruitment and selection procedure, performance management system, employee involvement, and training are significantly associated with improved employee turnover, productivity, and corporate financial performance. An empirical study by Jaselskis and Ashley (1991) illustrated that team turnover rate as a resource allocation strategy could predict overall project performance, schedule performance, and budget performance.

The U.S. Department of Energy (2010) published a "Staffing Guide for Project Management" of resources, allowing for a provision of the right personnel with the right skills to plan, direct, and oversee DOE projects. The staffing guide was based on several characteristics, and the staffing model provides a way to calculate the staffing needs for each functional unit. The agency stated that projects that are effectively planned and executed with an adequate level of oversight will be successful. At the contract administration functional area, the DOE staffing model could be used to identify the number of people needed to manage the functional area considering the key characteristics as outlined in Table 3.

Table 3 DOE Workload-Based Staffing Algorithm

Steps for Developing Project Staffing
Step 1: Establish the Project's Unadjusted Staffing
$\underline{PV}$ = PS (Unadjusted)
PF
Step 2: Adjust the Project's Staffing Based on Project Characteristics
PS (Unadjusted) + PS (Unadjusted) (PT + PC + PE + PP + RI + EI + PU + CT) = PS (Adjusted)
Step 3: Allocate the Project's Adjusted Project Staffing to Contract and
Project Management Functions
PS (Adjusted) x FAP = RS
Variable Acronyms
PV = Project Value (Annual Value of work to be executed by the project)
PF = Productivity Factor
PS = Project Staffing
PT = Project Type (a Factor Value)
PC = Project Complexity (a Factor Value)
PE = Project Execution (a Factor Value)
PP = Project Phase (a Factor Value)
RI = Regulatory Involvement (a Factor Value)
EI = External Influence (a Factor Value)
PU = Project Uniqueness (a Factor Value)
CT = Contract Type (a Factor Value)
FAP = Functional Area Percentages (recommended percentage staffing distribution by function depending on the project phase.)
RS = Recommended Staffing for Associated Functional Area
The productivity factor (PF) is defined as a reasonable amount of project dollars that a full time equivalent (FTE) can effectively manage in a given fiscal year, and is in terms of million dollars per FTE (M/FTE).

Having the experts involved at the onset of a problem makes a huge difference in management. General contractors should identify those experts, and they should be available for immediate assignment to resolve disputes and claims. Resolving change involves an enormous amount of documentation to prove entitlement. The situation is even made worse when the right people are not available to start the process of evaluating the conditions and documentation and getting on with negotiations. Ren, Anumba, and Ugwu (2000) stated that when the contracting parties are late in getting involved in addressing issues, the result is inefficient resolution of claims. The authors recommended the use of a multi-agent system (MAS), based on reasoning model and negotiation mechanisms. The reasoning model discourages self-interest negotiation and advocates for negotiation based on overall project interest and willingness to compromise. The negotiation mechanism relies on early involvement of all parties concerned, and that negotiation is a continuous process, and compromises have to be maintained or else the negotiation becomes one sided and breaks down.

Contracting involves various contracting parties, and each party is expected to meet their part of the contract. Just as a general contractor is expected to establish and maintain a contract administration department, it is equally important for owners to have a functional unit responsible for contract administration. There is a need for owners to have a contract administration department that can effectively manage general contractors' performance. A general contractors' ability to meet contract administration would mean that both owners and general contractors are getting the value for their investment.

Resolutions of disputes require high amounts of management oversight and man-hours, and this could explain why it is important to manage disputes using a procedure that reduces the amount of resources used. Cheung, Tam, and Harris (2000) upheld that construction dispute resolution has remained a big concern due to the growing number of resources required to resolve disputes, and early management involvement was cited as one of the key variables required for effective resolution of disputes.

Contract administration starts from the point a contract is executed and it is important that resources are assigned immediately when the contract is signed in order to know who will manage the changes that will occur and get them involved in the process. This strategy could differentiate organizations that have contract administration as a core process and those that do not. Projects are dynamic and experience has shown that change happens. General contractors should assign contract administration teams to the project at the start of the project instead of waiting until problems are encountered before assigning contract administrators to the project. Quinn (2005) found that organizations that have procurement as a core competency generate 133 percent greater return on their investment in procurement than average companies. Quinn also found that procurement operation costs are 20 percent less for world-class organizations than they are for typical companies. Avery (2004) reported that a research by the Aberdeen Group estimated that a $1 billion company on a $400 million contract loses $18 million every year due to the lack of proper contract management.

Competency of Contract Administrators

According to Project Management Institute- PMI (2002), project success has two key variables: managers with the right competency and organization with maturity, environment, structure and capability. A manager with the right competency and an organization that is not mature or capable will not generate successful performance, and vice versa.

Knowledge management involves the process of acquiring, creating, sharing, utilizing, and storing intellectual assets that are internal and external to a business environment (Kazi, 2004). The key to building knowledge-base is on creating the right environment to build and enhance competency of contract administrators. While knowledge is an understanding gained through experience or study, skills are abilities that transform knowledge into use. Contract administrators should be knowledgeable in contract terms, conditions and specification, because no other risk mitigation measure is as important as knowing the contract. Molly (2007) stated that the first step in effective change management is to know what is in the contract. Havers,

O'brien, and Stubbs (1996) posit that disputes are the result of conditions where parties disagree on the existence of an issue, which party was at fault, its impact, when it became an issue or what solution to take. A dispute is a disagreement between the contracting parties on a contract issue. Quite often, disputes arise because of arguments emanating from one party not having full knowledge of the contract conditions and requirements. Most contract specifications are structured to include three main parts: the bidding and contractual documents and forms, the conditions of the contract, and the technical specifications (Fisk & Reynolds, 2010). According to Katz (n.d), risk management begins with the contract and prudent parties are expected to be knowledgeable about the contract conditions and what is required to meet those conditions.

Contract administrators should be knowledgeable of the impact of change on cost and time, and how to estimate changes. For example, knowledge of "Measured Mile" and its use and application to change order pricing should be well understood by practitioners. The need for contract administrators to have good knowledge in the principles of measured mile was captured by Portilla (2010), where the author noted that construction disputes could be avoided if practitioners understand the concept of measured mile and implement it appropriately. Measured mile is a change evaluation method for lost labor productivity, and compares an impacted item of work with an unimpacted similar item of work, where both have the same baseline conditions. The aim is to identify contributing factors that are not found in the baseline conditions, and evaluate them for their economic impact on the project. Fínez (2008) used three main criteria to measure entrepreneur readiness towards a particular business idea, as in their level of competency, and completeness of their business plan. The competency criteria included attitude, aptitude, and capacity. The same concept can be used to evaluate contract administrators' competency. According to Fínez (2008), aptitude metrics include knowledge of sector, technical background and job experience, and attitude metrics include motivation, behavior and mentality, while capacity metrics include leadership, management and teamwork. Fínez (2008) mentioned that the importance of combining competency and a business plan is that one may have the competency in one business idea, but may be inadequate for another. This leads to the conclusion that a well-structured contract administration system is needed to allow application of the right competences.

Some other measures that could be used to evaluate competency of contract administrators may relate to the ability to prepare cost proposals with correct markup, correct documentation, correct labor rates, correct equipment rates and correct production rates. Also, making sure that the project baseline schedule or update schedules are current, and approved for use in time impact analysis (TIA). Li (2005a) points out that a baseline schedule is a must-have tool for project control. A study by Griffith (2006) found that projects with well-defined and detailed schedules are correlated with successful performance in cost and schedule. This is because a well-defined schedule provides a basis to manage and control changes on a project. Other criteria for evaluating contract administrators' competency include maintenance of daily reports specific to each operation. Portilla (2010) pointed out that daily reports are important documentations to record construction operation as well as record changes when they occur. The importance of accurately representing and forecasting schedule activities based on analysis of past production (historical data) was examined in a study by Li (2005b). The author found that without correct analysis, and consideration of the productivity, the schedule forecast will most likely be flawed. Contract administrators should be knowledgeable about different TIA methods and specifically of interest is the use and application of contemporaneous TIA. The contemporaneous TIA method evaluates the differences in project completion date by taking snapshots of the project before and after a major impact has occurred (Mohan & Al-Gahtani, 2006). In claims analysis, the effective implementation of Phased Root Cause Analysis (PRCA), as noted by Sandlin, Sapple, and Gautreaux (2004), depends on contemporaneous tracking and documentation of cost and schedule variance. Contract administrators should know the principles behind the use of contemporaneous analysis to provide a clear indication of what happened, what was the cause, and which party was responsible. Competency of contract administrators could also be evaluated on the basis that submittal of the baseline schedule or update schedules were not delayed for one (1) month or more beyond their due date. Also, competency could be evaluated on the basis that contract issues are addressed weekly, or on the basis that a project specific contract administration plans are available on site and in use by the general contractor.

In managing risk associated with change to contract, a risk register that has been approved by management should be contemporaneously updated to reflect current conditions. According to Kaliprasad (2006c), risks are analyzed and managed differently as they move from uncertainty to certainty, which is when they become an issue. "An issue is an event that is certain to occur (or may have already occurred)" (Kaliprasad (2006c, p28). Good risk management measures will indicate if subcontractors are prompt at sending notification of change, and pricing of change orders and also, if a general contractor is prompt at processing and requesting payment for actual work completed on executed change orders. According to Schleifer (1990), all contracts spell out payment provisions, yet many general contractors fail to comply with these provisions and end up not being paid on time. Other factors to assess knowledge of contract administrators include if the general contractor is prompt at meeting timely notification of changes, and notice of change is sent to the owner as soon as the general contractor has knowledge of an issue. In practice, the duty to notify is the responsibility of the general contractor, and this is a very important aspect of change management. The objective is to allow the owner to mitigate the risk as soon as the owner is aware. According to Levin (1998), for the contractor to preserve their rights for equitable compensation in time and costs, the owner must be formally notified for changes or claims. Levin, posited that notification allows both parties to verify conditions, assemble facts, and resolve the problem. Failure to notify the owner of problem, puts the burden of responsibility to resolve the problem on the general contractor, and this is why the importance of notification cannot be overemphasized, (Levin, 1998). Approval of a monthly update schedule should be current and no more than one month behind, and the project schedule should contain a work breakdown structure (WBS) that makes it easy to filter, organize and isolate activities. According to Rad and Cioffi (2004), "a good WBS anchors a project's plans and improves planning, estimating, monitoring, and controlling" (p. 31).

According to Lozon and Jergeas (2008), the three main areas that create construction disputes include project uncertainties, problems resulting from how well the design and construction processes meet functional requirements, and people issues. The increasing number of disputes and claims experts readily available illustrates a gap created when general contractors lack in-house experts to manage change, disputes and claims. However, there are also downsides to the use of consultants in managing change. It is less expensive to use in-house experts instead of an outside consultant to prepare cost proposal, TIA, and claims. Cheung, Tam, and Harris (2000) found that one of the variables that hinders the resolution of disputes is the use of outside consultants. The authors stated that the use of external claims advisors may lead to the positioning of rights, and thereby hinder parties from working towards the same objectives. From a different point of view, Levin (1998) acknowledged that using consultants can be costly, but the author also recognized that consultants are most economical in areas where the contractor's expertise is limited. Use of outside consultants would indicate that additional costs may be added to the cost proposal due to the cost of hiring a consultant. One way to gain the same level of knowledge as the consultant is to have the general contractor's contract administration managers join one of the professional associations (e.g. Project Management Institute-PMI or Association for Advancement of Cost Engineering International-AACEI) or become certified by one of these professional associations. Most consultants belong to these associations, and contract administration managers can benefit from literature, best practices and standards published by these associations. California Department of Transportation (Caltrans) has recognized the need for their project managers to gain knowledge of the principles and practices of project management. In this regard, Caltrans has teamed up with several Universities such as California State University Sacramento, College of Continuing Education to offer project management certification based on a standard written by PMI. Caltrans has also written "Caltrans Project Management Handbook" based on PMI's Guide to Project Management Body of Knowledge (PMBOK), and the 5Ed was released in 2007 (Caltrans, 2007).

The need for the general contractor to provide relevant and applicable training to employees involved in the contract administration process cannot be overstated because of what is at stake in the contract. According to the U.S. Merit Systems Protection Board (2005), the volume of federal contract spending was $328 billion in FY2004, which is 87 percent higher when compared to the FY1997 spending of approximately $175 billion. The U.S. Merit Systems Protection Board report stated that the increasing value of expenditure requires efficient and effective management of the contracting officers involved in these contracts. The U.S. Merit Systems Protection Board (2005) study of 10 federal agencies that accounted for 90 percent of the U. S. Government's contract expenditure, found that key items that must be addressed in meeting contract objectives include formal delegation of authority, improved training, and strategic management of the workforce. According to Kaliprasad (2006b), an organization will have the capability to adapt to changes if it is involved in constant learning, active listening, understanding, and aligning with the current issues that affect all stakeholders.

Just like the saying goes, a picture is worth a thousand words, visual charting, graphing or other visual representation should be used when presenting the impact of changes in the contract. Visual flowcharts could also be used to complement the understanding of the contract provisions and recovery steps. Sturgill and Vorster (2006) worked with the Virginia Department of Transportation (VDOT) to develop visual flowcharts as another form of representing contract conditions. The idea being that poor contract practices leading to disputes are a result of difficulty in comprehending contractual conditions in text form. Sturgill and Voster developed a simple framework to organize contractual conditions using a visual aid in the form of flowcharts. The research found that the use of a flowchart provided a better depiction of contract conditions and the required actions to be taken.

Project data are everywhere, but they are only informative when they are rendered in a format that allows users to make sense of them quickly and act on them. The use of visual representation is important in painting a clear picture of an issue. The work of Yau (2011) and Few (2009) provided concepts, principles, and practices on how to use visualization methods to show variances, patterns, and trends of events or issues over time, show how events or issues can be groups by various categories, and how events and issues can be represented to show relationships or make comparisons. The goal is to represent information visually, and allow practitioners to make the right decision when faced with an issue or a problem.

TRENDS AND ISSUES THAT MAY INFLUENCE THE FUTURE OF CONTRACT ADMINISTRATION PERFORMANCE

System forecast may allow managers the ability to reduce project risk by providing a view of what is new or what is going on within the environment, and thereby reduce the level of unknowns. Understanding the trends in a system should help managers see changing environment, and prepare for their impact on project objectives.

In addition to the practices discussed in the preceding section and their relationship to contract administration performance, the trends and issues discussed in this section are also expected to influence general contractors' contract administration performance. Organizations forecast future trends in order to take advantage of future opportunities, and mitigate the risk associated with the future events. Predicting the outcome of the future may be one of the most important and difficult functions of a manager. Effective managers may be those that opt for a proactive approach instead of a reactive approach to management of future outcomes. The ability to predict or anticipate future trends, opportunities, and impact should be a competitive edge.

Contract Delivery Methods

The National Academies (2007) identified that one of the best practices to reduce construction cost associated with disputes is the use of an alternative contract delivery method such as "best value." The recommendation was the use of "best value" approach in bid selection, as opposed to a low bid process. A good example is the use of design-build method, where the contractor-designer company takes the risk of design and could no longer go against the owner for omissions and error resulting from the design. However on any given project, changes to the contract may be the result of several sources, and while contract design issues may no longer be one of them, it is still important that general contractors are prepared to manage all other resulting changes.

Right Sizing of Project Partners and Workforce

The shortages of a qualified workforce will continue for both craft and non-craft. It is expected that with integrated information systems, the future will require fewer project teams with cross functional training for optimal and efficient staffing and management of projects.

In reference to the issue of sustainability, Bourdeau (1999) stated that, in the future, it is expected that professional compartmentalization will become a thing of the past, and will be replaced by the use of multiskilled, multidisciplinary managers and operators. Sustainability, lean construction and appropriate technology are all aimed at reducing waste, preserving resources and supporting socio-economic systems, and all three concepts have a place in the practice of contract administration through availability and utilization of the right processes, tools, techniques and infrastructure.

New and Emerging Technologies

New technologies such as Building Information Modeling (BIM) will present challenges and changes to how construction operations are conducted, and some of which might be cultural and organizational changes that will impact design, and management of contract. According to Ofori-Boadu, Okere, and Kim (2010), implementation of BIM across an organization will require a shift in how business is conducted, and practitioners will have to learn new skillset to manage and administer projects.

PERFORMANCE MEASUREMENT

While contract administration performance may not be measured directly, a good understanding of how to measure contract administration performance is required in order to capture key indicators at all levels.

Performance management relates to controls in response to "feedback or information on activities with respect to meeting customer expectations and strategic objectives" (Wegelius-Lehtonen, 2001, p. 108). According to Davis, Aquilano, and Chase (2002), the types of performance measurement include productivity (output/input), capacity (actual output/design capacity), quality, speed of delivery, reliability and process velocity (total throughput time/value added time).

What models and methods could be used as a guide for measuring contract administration performance? Imagine two project teams working on a similar project design. Both teams' objectives are to meet budget and time. Team A spends much time planning out the project, and finishes on budget, meets requirements, but completes the project one week late.

The performance of Team A could be seen as good, because the team achieved the budget objective, but they were one week late. Team B also went on to build a similar project. However, rather than planning out the project before starting, they planned as they went on, and had numerous rework, safety issues, compliance issues, false claims. As luck would have it, Team B finished on budget, and finished one week early. For achieving both project objectives, Team B may be considered to have been more effective, although they had more performance issues than Team A. Unfortunately this type of evaluation is subjective, and fails to help both teams to understand areas of improvement. According to Rad (2003), the current methods of performance measurement are subjective, inconsistent and not transparent; resulting in situations where a project fails in several areas yet is evaluated as being successful. On other occasions, the team may consider their performance a success, yet the stakeholder's evaluation may indicate that the team failed to meet objectives.

A true measure will be consistent from project to project and help remove false evaluations of performance, as well as provide a proactive tool for practitioners to take corrective steps. The mantra is "what gets measured gets improved" but the key is in having the right measurement tool. Several federal agencies have developed a composite measure of contractors' performance (Appendix E). Currently the contractor performance assessment is used as an evaluation tool for selection of contractors on "best value" projects. The items in the contractor performance assessment form provide for a composite evaluation of contractor performance. Perkins (2008) argued on the validity and legal problems arising from the use of the contractor performance assessment form on federal projects. The author contended that the method is unfair and biased. A reliable and valid assessment tool is more relevant now than ever before as more federal agencies move away from a request for bid (lowest bidder) to a request for proposal (best value). This study addressed this problem by providing a tool to evaluate a contractor performance at the contract administration functional level. The State of Utah Department of Transportation has developed a contractor performance rating form for evaluating the overall performance of a contractor on a project (http://www.udot.utah.gov/main/uconowner. gf?n=200510270725131). Also the State of Kentucky Department of Highways has a contractor's performance report form (http://transportation.ky.gov/Organizational-esources/Forms/TC%2014-19.pdf). While these performance evaluation forms capture some of the key practices that underlie contract administration practices, they do not correlate the practices to appropriate contract administration performance, and could not be used for predicting contract administration performance of general contractors on state DOT projects in the U.S.

It has been argued that project performance should be measured using metrics that could localize and identify the underlying problem instead of the current way of using change in cost, and change in time, that only takes a high level view of the project (Gransberg & Villarreal Builrago, 2002). Shenhar, Levy, and Dvir (1997) viewed project success from the perspective of its contribution to organizational and societal progress. They found that project success can be measured by using four groups of indicators that include project efficiency with respect to how project processes were managed, impact on customer objectives and requirements, business or commercial success to the organization, and preparation of organizational infrastructure for the future. The same model can also be extended in evaluating the success of a functional unit such as contract administration.

Korde, Li, and Russell (2005), conducted an extensive research on some of the studies completed over the last 20 years in the area of construction performance measures and critical variables that were thought to be the drivers and cause of performance variability. In their work, Korde, Li, and Russell looked at 122 academic papers on causal models published in the last 20 years in various construction related journals and conferences, and they found that the main performance metrics in these papers centered on production, cost, quality, time, safety, project success and others. Also, the levels of analysis of the studies were categorized into three areas which included project level, group level and activity level. However, none of these studies were on performance metrics related to general contractors' ability to meet contract administration objectives consistently on federal and state DOT projects in the U.S. Also, the levels of analysis identified in these studies were mostly at the project level and a few of them were at the group level which is aligned with the functional areas. This study was centered on contract administration as one of the many functional areas that contribute to overall project success or failure. A contract administration system capable of meeting contract administration objectives must have 1) foundational components, elements, processes, tools, techniques and infrastructure, and 2) efficient and effective utilization of these items in order to meet or exceed contract administration objectives. The goal of contract administration is to meet objectives that include compliance with terms and conditions (T&Cs), owner relations, change management, dispute resolution, and claims resolution. To achieve these objectives, a general contractor must provide and utilize the right practices.

Measuring Processes and Operational Performance

A measure of process captures the variations in the environment, conditions, the 4M (machine, method, material and manpower), and allows for control and prediction of outcome. The problem is in finding the right indicators to measure the health state of the process. According to Brown (1996), consistency is about using the right process in the right way and having the right metrics. Process measurement starts with a well-defined requirement of the end product or service, and then determines the impact of each process on the end product or service. Excellence is about control and monitoring of organizational processes to generate product and services consistently. In order to control and monitor performance at all levels, an organization must monitor and control the conditions, environment, inputs, processes, outputs, feedback, and outcome (Brown, 1996).

Three Level Performance Measurement

Rummler, Ramias, and Rummler (2009) argued that a sustainable organization must plan, design and manage its performance at three levels. Rummler and Brache (1995) provided a structure to measure a system performance based on a holistic approach which they believed represents an anatomy of performance. The three levels include the organization level, the process level, and the job performer level. In addition to the levels, there are also a) the goal needs that determine specific standards and objectives at each level, b) the design needs at each level that reflect structures (who, what, where, when, why, and how) in place that enables the goals to be achieved and, c) the management needs at each level that ensures that goals are current and are being achieved

CHAPTER 3

RESEARCH METHODOLOGY: OBJECTIVE EVALUATION OF WHAT IS GOING ON

Following the established procedure of a quantitative correlational research, what method shall be used to accomplish the objectives of this study as pointed out in Chapter 1, and why? In this section, data are gathered to a) confirm or refute the proposition on performance factors that are related to general contractors' ability to meet contract administration objectives on federal and state DOT projects in the U.S., and b) build a model to identify, evaluate, predict and control contract administration performance of general contractors on federal and state DOT projects in the U.S. This study provided for a test of the hypotheses, and a development of a model for use in solving contract administration problems of general contractors' ability to meet contract administration objectives consistently on federal and state DOT projects in the U.S.

Based on the literature review on what factors are responsible for general contractors' ability to meet contract administration objectives, the goal was to verify if the hypotheses aptly describe how contractors practice contract administration. To meet this goal, the study gathered information needed to answer the question posed, accepted or rejected the hypotheses, and thereby help solve the problem of contract administration performance.

As a guide, this study followed Li, Korde, and Russell's (2005) recommendation on how to enhance the industry's acceptance of construction research that are related to project performance. Their recommendations are:

- Make use of existing and already available day-to-day data.

- Explanatory methods should be simple enough for practitioners to implement using standard office applications without need for software or customizations.

- The techniques required to generate the report should be readily compatible with the skill set of technically trained construction personnel without the help of consultants.

- Users should be able to formulate and update their own causal models of performance based on their experience.

- Causal model should be based on fundamental relationships where they exist.

POPULATION AND SAMPLE FRAME

The population of interest for this study was general contractors that were working on federal or state DOT projects in the U.S. at the time of this study. The population of interest reflects general contractors working on federal and state DOT contracts in the U.S. with contract administration provisions written into their contracts.

Thirty-six state DOTs were contacted to participate in this study, and only twenty of the state DOTs participated. The general contractors on these states DOT projects represented the sample frame for this study. The subjects being evaluated were the general contractors working on one of these state DOT projects where data were collected to assess and investigate their contract administration practices. The projects chosen for this survey were those that are one year or more into construction and have encountered some changes. This was significant because projects that are a year or more into construction are more likely to have encountered a sizable amount of changes, and would have data available on how long it takes to resolve change orders from when the change was encountered to when change order was executed.

The project delivery method was also taken into consideration for selecting projects for this study. This study was centered on design-bid-build delivery method, and only projects that fall under this group were chosen for this study. Design-bid-build still remains the dominant method of project delivery method for state DOTs. Major capital improvement projects constructed annually by various state DOTs range from the high 600 to the low 100 annually. For example:

- The State of California Department of Transportation - a review of ongoing contracts indicates that the department averages over 600 major capital improvement projects annually - http://www.dot.ca.gov/hq/construc/statement.html

- The State of Florida Department of Transportation - averages over 200 construction projects annually -

http://www.dot.state.fl.us/publicinformationoffice/moreDOT/majorprojects.shtm

SAMPLING METHOD AND SAMPLE SIZE

This study made use of the simple random sampling method, with an objective of selecting a representative sample from the population of interest in such a way that each subject had an equal chance of being selected. It was found to be too expensive and time consuming to gather information from the entire population of interest. Fortunately sampling provides a less expensive and less time consuming alternative. The use of the random sample was based on the idea that a random sample has the same characteristics and attributes as the population from where the sample was taken. However, it was important that the right sample size was obtained. Having the right sample size allowed for the statistical power required to detect associations and differences if they do exist. Not having the right sample size would mean that an association and difference may not be detected even if they do exist in the population, hence, the need for the right number of samples. The right statistical power which was the probability that the data gathered in an experiment, will be sufficient to correctly reject the null hypothesis, and avoid making a type II error (occurs when a statistical test incorrectly fails to reject the null hypothesis when it is false). The desired goal was to have a statistical test that rejects the null hypothesis when it is not true. This was in fact about significance of associations or differences that are affected by the number of cases in a sample. Relationship between two variables that is significant with larger sample size may be found not to be significant when a smaller sample size is used (Bryman & Cramer, 2011)

Miles and Shevlin (2001) suggested the use of power analysis for identifying sample size. Statistical power is the ability to find a relationship or difference in a study when a real relationship or difference exists. The power of a study is determined by three factors: the sample size, the alpha level, and the effect size, and power analysis allows researcher to compute the right sample size. Using charts from Miles and Shevlin, the appropriate sample size for this study was estimated to be 100, and this was based on alpha set at 0.05, medium effect size, power set at 0.8, and six (6) independent variables.

The criteria used in selecting projects included in the study were as follows:

1. project was a capital improvement project (major, non-recurring expenditures, such as roads and bridges)

2. project was one year or more into construction

3. project has encountered and executed some change orders

4. project delivery method was design-bid-build (competitively bid projects)

A contact person was identified for each participating state DOT to coordinate the administration of the survey. The contact person distributed the survey to the resident engineers that oversee the select projects.

On state DOT projects, the responsibilities of a resident engineer may include the following:

- Assign work, give instructions and make decisions on various engineering and contract administration issues.

- Make engineering decisions, and ensure compliance with plans and specifications.

- Monitor general contractor's work and progress.

- Review pay quantities and prepare daily reports.

- Review and approve contract change orders, progress and final pay estimates.

- Analyze and make decisions to resolve contract claims.

- Ensure that construction conforms to codes, standards and regulations.

- Coordinate resources, resolve various personnel related issues, and provide training and guidance to employees.

The respondents were asked to indicate which response option indicated what practices were in place and in use. The reason to use the resident engineers instead of the general contractors' representatives to complete the questionnaire stems from the view that it will reduce self-evaluation, social desirability (tendency to present oneself in a way that makes one look good), and bias from the general contractors. It was also expected that the response rate would be higher, being that the general contractors may not be motivated or obligated to complete the questionnaire. Finally, the approach allowed for the gathering of data in a normal setting without influencing the results of the study as the general contractors are likely to change their practices when they know that they are being evaluated on some criteria.

WEB-BASED QUESTIONNAIRE ADMINISTRATION

Following the study by Spitz, Niles, and Adler (2006) on the current state of practice in web-based surveying, this study collected data from respondents using a web-based administration. Some of the advantages of web-based administration are that it allows for immediate validation of the response, provides real time support to the respondents, monitors response rate, manages the delivery of survey in real time, and generates some basic descriptive statistics as data are collected. Qualtrics was used for survey design and hosting of this study, and it was administered through Indiana State University. To fully take advantage of the web-based survey, a contact person was identified from each state DOT that participated in the study. The contact person coordinated and approved the web-based distribution of the questionnaire to the resident engineers. Using Qualtrics, the web-based distribution email to the resident engineers originated from the agency's contact person. The idea was to avoid direct mailing of the questionnaire, as they were likely to be neglected, blocked, or filtered if the questionnaire email came from someone outside the agency.

UNIT OF ANALYSIS

Unit of analysis can be defined as the context of analysis. Unit of analysis is the individual, group, or organization described by a variable or set of variables (Mohr, 1990). According to Mohr, in relational studies, all the variables must be descriptors of the same unit of analysis. In order to generalize research findings, the generalization has to match the unit of analysis used in the research. In this study, the dependent variable was about the general contractors' ability to meet contract administration objectives, while the independent variables were about practices or actions of the general contractors' contract administration on federal and state DOT projects, and speak to the performance of the general contractors on federal and state DOT projects. The interest of this study was on how project conditions and practices relate to general contractors' ability to meet contract administration objectives on federal and state DOT projects in the U.S.

HYPOTHESES TESTING

The study hypotheses were based on research questions and were simply stated to reflect one dependent variable to one independent variable. Also they were specifically stated without ambiguity on the variables and the population of interest. With respect to a population of interest consisting of general contractors working on federal and state DOT projects, the tentative theories of this study included the following:

P1 – Ha: Management attitude towards contract risks is positively correlated to contract administration performance

P2 – Ha: Contract provisions for mitigating contract risks are positively correlated to contract administration performance

P3 – Ha: Stability of scope definition is positively correlated to contract administration performance

P4 – Ha: Contract administration infrastructure is positively correlated to contract administration performance

P5 – Ha: Resource allocation strategy is positively correlated to contract administration performance

P6 – Ha: Contract administrators' competency is positively correlated to contract administration performance

In this study, a quantitative correlational research method was designed to determine whether a positive relationship exists in the population of interest.

Correlation analysis is concerned with the degree of relationship between two variables, while regression analysis is concerned about making prediction of one variable given other variable(s). To answer the research question posed, statistical analysis (descriptive and inferential) was conducted on the data collected from the respondents and used to evaluate if a positive relationship exists and to what extent. In other words, test the hypotheses of this study. It is important to note that the test of hypothesis serves to corroborate or refute the hypothesis based on the weight of supporting evidence present in the data, but a theory can never be proved to be true. Also, the statistical analysis is capable of showing the level of association between the variables. To achieve these objectives, the study computed a correlation analysis to identify practices that relate to general contractors' ability to meet contract administration objectives consistently. According to DeVellis (2003), a high correlation indicates that individual scores from both variables occupy similar locations on their respective distribution. To develop the predictive model, the study computes a multiple regression analysis, which involves the determination of the degree of relationship in the patterns of variation of two or more variables. The ability to see patterns allows for predicting the environment, and this is based on the understanding that patterns do not form by chance.

The test of hypothesis associated with the statistical analysis serves to assess if the data is incompatible to the null hypothesis which states that a relationship does not exist. When the test of hypothesis shows that data are incompatible to the null hypothesis, this means that the relationship found was unlikely to be by chance, but due to some patterns that exist at least within the group studied. Theory should define relevant variables to include and irrelevant variables to exclude (model specification), while avoiding specification error due to use of wrong variables in a regression equation. Besides, a good theoretical underpinning provides a foundation for further research that must go on at a deeper level to uncover missing or omitted factors related to the phenomenon of interest. However, it was critical that the quality of the indicators was appropriate to capture the underlying variables of interest. According to Toole (2006), there have been cases where researchers spent months collecting data, only to find out that the statistical analysis were inconclusive due to poor indicators.

RELIABILITY AND VALIDITY OF INSTRUMENT

A questionnaire is an instrument that bridges the gap between abstract concepts and empirical indicators, and it has two key properties that include 1) reliability of an instrument to collect the same data consistently on repeated use, and 2) validity of an instrument to measure the concepts it is designed to measure.

Reliability of the questionnaire is concerned with making sure that the instrument is such that it is capable of collecting the same information when administered to the same population of interest repeatedly at any time. It is about consistent measurement, which is in fact about the accuracy of measurement. Reliability is measured in different ways as they relate to the degree to which a score is stable and consistent when measured at different times (test-retest reliability), in different ways (parallel-forms and alternate-forms), or with different items within the same scale (internal consistency).

The test-retest method was used for this study to evaluate reliability of the instrument, and the data from the two stage expert panel review of the instrument were available to measure the coefficient of reliability using correlation. The result of this test was reported under the section on "instrument and questionnaire design." Another method that could be used to measure reliability in this study is internal consistency, which measures how well items on a scale "fit" together to measure the same construct. Internal consistency is statistically estimated by Cronbach's Alpha.

Validity, on the other hand, is concerned about making sure that the instrument measures the right indicators that relate to contractors' ability to meet contract administration objectives within the state DOT environment. To answer the question "valid for what purpose" the following attributes of a valid instrument were examined:

Face validity- respondents saw the instrument as fair, because it is made up of items that assess contract administration capability of a general contractor on state DOT projects, and are typically designed to assess general contractors' use of the right practices.

Content validity- operationalized items in the instrument represented variables found in the body of knowledge of contract administration and are associated with the phenomenon of interest.

Criterion- related validity (predictive validity)- based on empirical results, it was expected that the instrument total item score obtained by evaluating general contractors' practices (independent variables) relate highly to the score obtained from operationalizing general contractors' ability to meet contract administration objectives (dependent variable). Criterion-related validity is measured using a correlation coefficient.

Construct validity- that the variables were operationalized correctly based on theoretical body of knowledge on the phenomenon of interest. Items in the instrument were carefully defined to provide scores that assess the capability of general contractors based on their practices as well as corresponding ability to meet contract administration objectives.

DEPENDENT AND INDEPENDENT VARIABLES

The dependent variable:

Contract administration performance was defined by the ability to meet contract administration objectives as they relate to compliance with terms and conditions (T&Cs), change management, owner relations, dispute resolution and claims resolution.

The independent variables:

1. Management attitude towards contract risks- relates to management involvement, perceptions, and behavior on issues of contract risks.

2. Contract provisions for mitigating contract risks- relates to contract provisions in place to address various issues and conditions that may be encountered on a project.

3. Stability of scope definition- relates to the rate and amount of scope changes, and provision in place to quickly resolve design revisions and changes.

4. Contract administration infrastructure- relates to the resources, tools and techniques in place to manage project documents, costs, and schedule.

5. Resource allocation strategy- relates to availability and timely assignment of project personnel.

6. Competency of contract administrators- relates to knowledge, skills, abilities, and traits of a contract administrator.

OPERATIONALIZATION

Force, mass, acceleration, distance, pressure, stress, moment, energy, for example, are easily measured, and well understood concepts within the construction environment; however, most social science phenomenon are not easily measured. In order to evaluate the relationship between the key constructs of this study, the constructs were further defined in terms of measureable items. The need to operationalize the constructs (latent variables) allowed for statistical evaluation and quantification of the extent to which items of the constructs were interrelated and predictive. A collection of items (effect indicators) that share a common cause, and are meant to measure the existence of variables not readily observable by direct means is referred to as a scale (DeVellis, 2003). Since relationships of some theoretical constructs or phenomena are not directly observable, a scale must be developed to operationalize the variables to observable proxies and allow for quantitative evaluation of relationships among variables of interest.

In this study, contract administration was defined as the process of managing all elements of a contract phases with the goal of meeting compliance, managing changes, maintaining good owner relations, resolving disputes, resolving claims and avoiding litigation. The research questions were formulated specifically to define what the study aimed to learn and understand, and through literature review, tentative answers were generated for the questions raised. However, in order to reach a conclusion on the tentative answers, data must be gathered and statistically determine whether and to what extent relationships exist.

It then becomes important to operationalize the key variables of this study, where the dependent variable was ability to meet contract administration objectives as they related to meeting compliance, managing changes, maintaining good owner relations, resolving disputes, resolving claims and avoiding litigation. The independent variables related to factors that include management attitude towards contract risks, contract provisions for mitigating contract risks, stability of scope definition, contract administration infrastructure, resource allocation strategy, and competency of contract administrators. For example, a study could be based on operational definition of a "loving family" as the number of times per week a family eats dinner together.

While other studies could define a loving family differently, it is important that the operational definition is one that is measurable, clear, apply to the context of interest, well understood, and can be replicated by others. As pointed our earlier, it is critical that the quality of the indicators is appropriate to capture the underlying variables of interest, because poor indicators contain random noise that may mask valid statistical relationship and may render the statistical analysis inconclusive. Appendix H provides a view on how the variables were operationalized.

Operationalizing the Dependent Variable

According to Cox (1997), successful management of change should be measured by how quickly the issues are resolved to the benefit of all parties.

In this study, the dependent variable was operationalized (describes how the construct was measured and reduced to a number) based on the understanding that a company that achieves contract administration objective consistently is one that attains lower cycle time from encountering change to reaching agreement and executing a change order. A study by Walker (1995) found that time performance is affected by four factors that include management effectiveness, cooperative construction environment, experience of client, and scope complexity. The time spent in responding to problems and getting them resolved could be an indication of how well a general contractor is doing in meeting compliance, maintaining good owner relations, managing changes, resolving disputes, resolving claims and avoiding litigation. When issues are quickly resolved, it allows organizations to reassign resources to other tasks, and by doing so, time, money and relationships are saved. According to Fisk and Reynolds (2010), claims should be handled promptly. The authors stated that "The second worst way to handle claims is to ignore them; the worst way is to allow them to go to litigation" (p. 173). Some of the benefits of partnering as reported by Caltrans is that claims are mitigated and resolved on time, and total costs are reduced (Caltrans 2008). The basis for operationalizing the dependent variable using cycle time is that quality of work is all about meeting objectives, increasing throughput (the rate at which a system achieves its goal), reducing rework, reducing waste, and getting it right the first time. However, when objectives are not met, the result is rework, waste of materials, time, and resources, or additional time and effort required to resolve the problem. If a process for handling contract changes and claims are executed right the first time, then there should be no rework. Schleifer (1990) argued for timely dispute resolution because delayed resolution presents a distraction from the normal business and a damp to team morale. Knowing how much time is spent resolving contract administration issues, provides the basis for evaluating general contractors' ability to meet contract administration objectives. The interesting thing about cycle time is that it reflects the time resulting from all three levels of performance identified by Rummler and Brache (1995) which include organizational, process, and performer level.

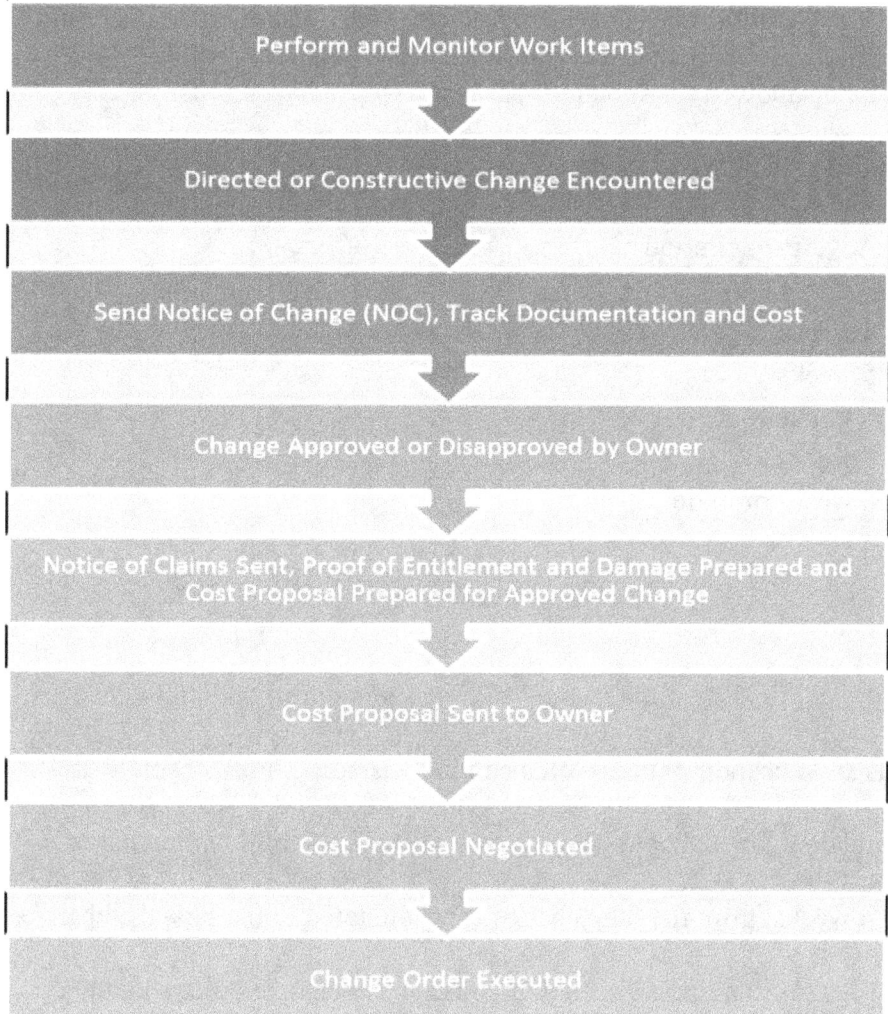

Figure 8. The Total Cycle Time of Change Management (created by Okere, O. G. to illustrate concept elements and the relationships of those elements)

Figure 8 illustrates the steps involved in cycle time for managing changes and issues. The steps are basic and include both value added and non-value added activities. These steps are similar to the steps involved in bidding projects. The steps involved in bidding a project include bid advertisement, bidder's inquiry, preparing bid package and bid opening. The bidding cycle starts from bid advertisement, to bid opening, and the cycle time is the duration given for bid submittal. This duration is indicated by owners, and in the case of Caltrans, some basic criteria are used to define the cycle time. A 2006 report by Caltrans on Major Construction Contract

Advertising Timeframes reported that the current system reflects predefined duration (Caltrans, 2006). Table 4 shows time frame allowed for submittal of bids based on the contract size.

Table 4 Construction Advertising Timeframe: Source Caltrans

Contract Size	Advertising Period
<$1 million	3 or 4 weeks*
$1-5 million	4 weeks
$5-15 million	5 weeks
$15-25 million	6 weeks
$25-50 million	7 weeks
>$50 million	8 weeks or more

Note: * Simple projects with 20 or less items or plan sheets and Safety 010 projects advertised for 3 weeks.

It has been stated by Cox (1997) that the amount of time spent in responding to changes and getting them resolved is an indication of how well a company is doing. From Table 4, it could be argued that a general contractor should be able to resolve a $1-5 million directed or constructive change within 4 weeks, from the time change is encountered to the time the change order is executed. Similarly, a $25-50 million directed or constructive change order should be resolved within 7 weeks of encountering change. The same concept could be used as the baseline for evaluating a general contractor's ability to meet contract administration objectives as they relate to compliance, managing changes, maintaining good owner relations, resolving disputes, resolving claims and avoiding litigation.

In this study, cycle time measured general contractors' ability to meet contract administration objectives, and there could be a strong relationship between cycle time and ability to meet objectives. When a general contractor manages changes to the extent that time from discovery of change to execution of change order is reduced, the owner sees the general contractor as being responsive. This level of responsiveness is also evident in all aspects of meeting contract administration objectives which include compliance to contract T&Cs, change management, owner relations, dispute resolution and claims resolution.

Operationalizing the Independent (Explanatory) Variables

The independent variables were operationalized by looking at contract administration processes and corresponding indicators based on the understanding that performances that provide for consistent achievement of contract administration objectives require the right processes and corresponding practices, and capabilities to be in place and in use. For this study, a questionnaire was developed to capture the practices that define each independent variable. The indicators that define each variable were validated and informed by current policies, and standard specifications of various state DOTs. These indicators were also validated by a panel of contract administration experts.

Questionnaires can be classified by their content. For example, some surveys focus on opinions while others are concerned with factual data. This questionnaire was aimed at collecting factual and objective information of the current state of practice. The advantage of collecting factual data is that the questionnaire is well suited for development of core knowledge areas, principles and best practices.

The study questionnaire items make up a homogeneous scale and reflect the factors (latent variables) underlying them. This is to say that each item is viewed as a measure of a latent variable, and should be an indicator of the construct of interest. These items were chosen from contract administration body of knowledge, and were also explicitly identified in the literature review section of this work. The key is that each item is an overt manifestation of a latent variable as the cause of variation in that item (DeVellis, 2003). Questionnaires with multiple items show that the phenomenon of interest could be influenced by any one of those items such that the number of variation is equal to the number of items on the questionnaire.

DERIVING THE QUANTITATIVE NUMBERS FOR THE DEPENDENT AND INDEPENDENT VARIABLES

The dependent variable was measured by the average length of time (in months) from discovery of change to change order execution.

The quantitative weight of each independent variable was generated by taking the score from each question per variable and collapsing them to make up the value of each independent variable. There were six key independent variables, and each of them have more than one item (indicators) that evaluate if the practice was in use or not in use- (the literature provided the understanding of what practices relate to effective performance).

The total score for each variable was the number of those practices (items) in use multiplied by one (1). One (1) represents the weight of each item when in use.

For example, "resource allocation strategy" was operationalized with a set of items, and each item response by a respondent was scored as one (1) for "YES" when the practice was in place and zero (0) for "NO" when the practice was not in place. Also, an item response was scored zero (0) for "Not Applicable" when in fact the item refers to a generally required state DOT practice. The total possible score for "resource allocation strategy" was one (1) multiplied by the number of items that operationalize "resource allocation strategy". According to DeVellis (2003), binary (yes, no) response options allow for adequate variation in scores when responses are aggregated for all responses that make up each variable being measured.

1. Dependent variable: Ability to meet contract administration objectives- measured by average cycle time (in months) from discovery of change to formal execution of change order

1. Independent variable: Management attitude towards contract risks- measured using several key indicators

2. Independent variable: Contract provisions for mitigating contract risks- measured using several key indicators

3. Independent variable: Stability of scope definition and requirements- measured using several key indicators

4. Independent variable: Contract administration infrastructure- measured using several key indicators

5. Independent variable: Resource allocation strategy- measured using several key indicators

6. Independent variable: Competency of contract administrators- measured using several key indicators

INSTRUMENT AND QUESTIONNAIRE DESIGN

In this study, a questionnaire as a data collection tool is a scientific instrument for gathering reliable and valid information for the purpose of analyzing if a positive relationship (correlation) exists, and creating a predictive (regression) model if the data showed that a high percentage of the dependent variable was explained by the independent variables (coefficient of determination).

A research instrument was developed to measure contract administration practices and their relationship to general contractors' ability to meet contract administration objective consistently. The instrument was a one-dimensional scale made up of various items measuring the same phenomenon- ability to meet contract administration objective consistently. The scale has equally weighted items for each independent variable item with scores of the scale to measure general contractors' ability to meet contract administration objectives. Also, the questionnaire subscales represented the latent variables of interest. To date, there is no method for measuring general contractors' ability to meet contract administration objectives on federal and state DOT projects in the U.S.

This instrument also provided for assessment of the level of maturity of an organization. According to Saxena (2008), maturity level of an organization, units, and functional area can be assessed by use of the right technology, infrastructure and capability that provides for optimal performance and excellence.

Two stages of validation were completed with a panel of experts using Qualtrics for web administration. The instrument used in the first round of validation contained 63 items and was distributed to 18 contract administration experts. The experts were asked to indicate if a question is valid for assessing contract administration performance of general contractors working on federal and state DOT projects in the U.S. The experts were also given the option to provide a suggestion for rewording questions for clarity. Nine (9) of the responses were returned, which amounted to a 50 percent response rate. Based on the responses received, it was inferred that some of the questions were not clear enough, and the questions were subsequently revised for clarity and a definition of a valid question was also provided. A valid question was defined as an item for which a change in the direction of practice will affect contract administration performance of general contractors working on federal and state DOT projects in the U.S. The second round of validation was then sent out to 8 experts, and there were 62 items in the instrument. Four (4) of the responses were returned and amounted to a 50 percent response rate.

Based on the responses received from the second round validation, the number of items in the

instrument was reduced to 59. Table 5 data is based on the scores obtained from the four (4)

experts that participated in both phases and resulted to a high reliability.

Table 5 Computing the Reliability of the Instrument

Expert Panel Members	Score at First Evaluation Phase (63 Items)	Score at Second Evaluation Phase (62 Items)
M.S	59	62
F.D	30	46
B.G	60	60
J.H	49	50
CORRELATION (Coefficient of Stability of Questionnaire)	0.920962332	

Prior to application for IRB approval, INDOT, NCDOT, NDOR and Caltrans were

contacted for participation. When Caltrans district 4 was contacted about the research, they

immediately assigned the "Office Chief, Functional Support" to work on this research project.

The "Office Chief, Functional Support" provided clarity to the questions and made sure that the

questions were aligned with current policies, and standard specifications. After the initial review,

two more rounds of reviews were conducted with the Office Chief, Functional Support, and all

the questions went through several iterations. Finally, an agreement was reached on 40 questions

that captured the six sub-scales.

With the survey question agreed on, IRB application was completed and IRB review

suggested that the general questions in the instrument be revised by removing one of the general questions to avoid re-identification, or tracing back which resident engineer completed which questionnaire. With the revision made based on IRB suggestion, the final instrument contained 40 questions on practices that relate to contract administration performance, 1 question that relates to how quickly changes are resolved, and 4 demographic questions.

CHAPTER 4

DATA ANALYSES AND FINDINGS: WHAT THE STUDY FOUND

In research, statistical analysis on data is made so that some interpretation or discussion can be made. There are two parts to a statistical analysis, the descriptive statistics and inferential statistics. In descriptive statistics the goal is to summarize and display data and then provide comments and interpretation. In inferential statistics, researchers go beyond describing data; they test if the study propositions aptly reflect what is going on.

Does correlation or regression imply causation? Neither correlation nor regression can indicate causation. While correlation is concerned with whether two variables co-vary, regression by contrast is concerned with the degree with which one variable changes in relation to another variable. Regression describes the dependence of variable Y on an independent variable X. Through observation and logical reasoning, an inference can be made based on characteristics of the observed experience as captured in the sample data. In which case, we can describe the characteristics of the observed experience; infer whether a relationship is probably causal, probably by chance, or that a relationship does not exist. In addition to making inference based on the facts collected, a test of regression model (where applicable) is used to evaluate and validate the effectiveness of the model at predicting outcomes given certain set conditions. This test is achieved by dividing the data in two parts. One part is used to fit the regression model, while the other part of the data is used to test how well the model predicts.

SURVEY DISTRIBUTION AND RESPONSE RATE

Thirty-six state DOTs were contacted with a request to provide 5 to 10 representative project to participate and complete the research survey. With 36 target state DOTs, the total expected project participation was computed at 180 respondents based on average of 5 projects from each of the 36 state DOTs. The number of projects that actually completed the survey was 86 at a response rate of 48 percent

DESCRIPTIVE STATISTICS

The descriptive statistics section aimed at describing data captured from the respondents. The data reflected the dimensions and profile of when, what, where, and how general contractors practiced contract administration within the state DOT environment in the U.S. Based on the questions in the instrument the descriptive statistics helped to summarize, and display answers to those questions based on data collected. Also, relevant comments and interpretations were provided to explain what the collected data meant, if the data was in line with what was expected, or why the data may have deviated from what was expected.

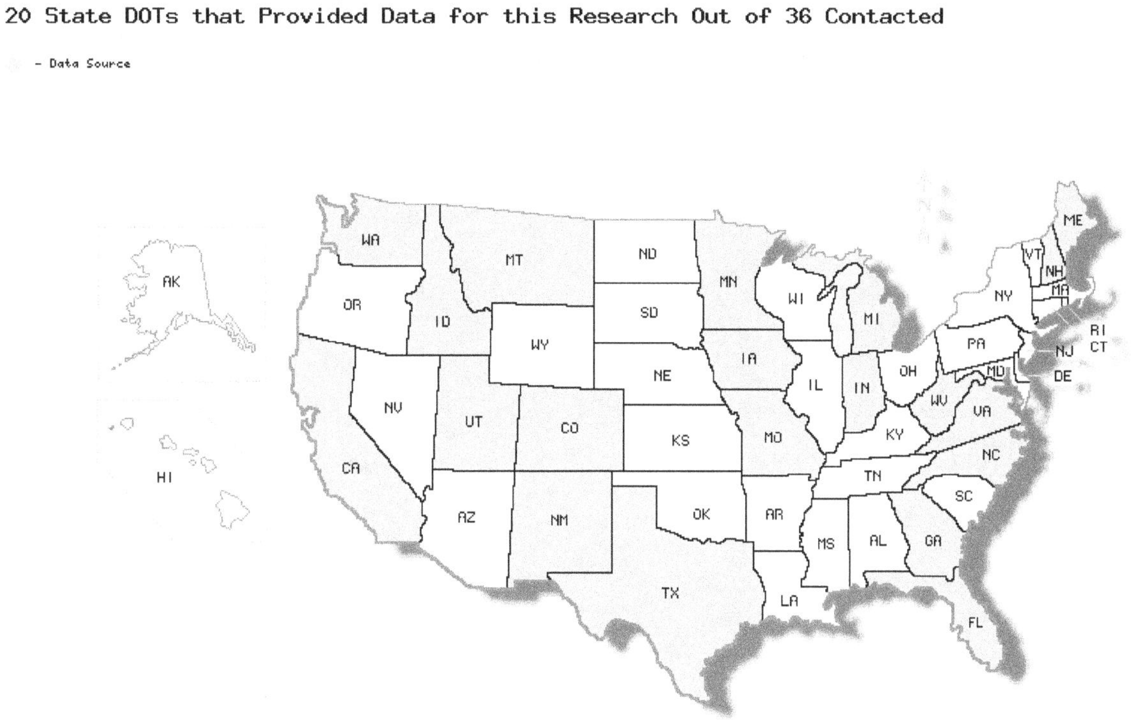

20 State DOTs that Provided Data for this Research Out of 36 Contacted

– Data Source

Figure 9. State DOTs that Provided Representative Samples for this Study

Figure 9 and Figure 10 show the spatial location of the state DOTs, and the number of projects that participated in this research, and it can be seen that data for this research has a good coverage of participation from state DOTs. The projects were randomly selected by representatives of the state DOT that participated in this study.

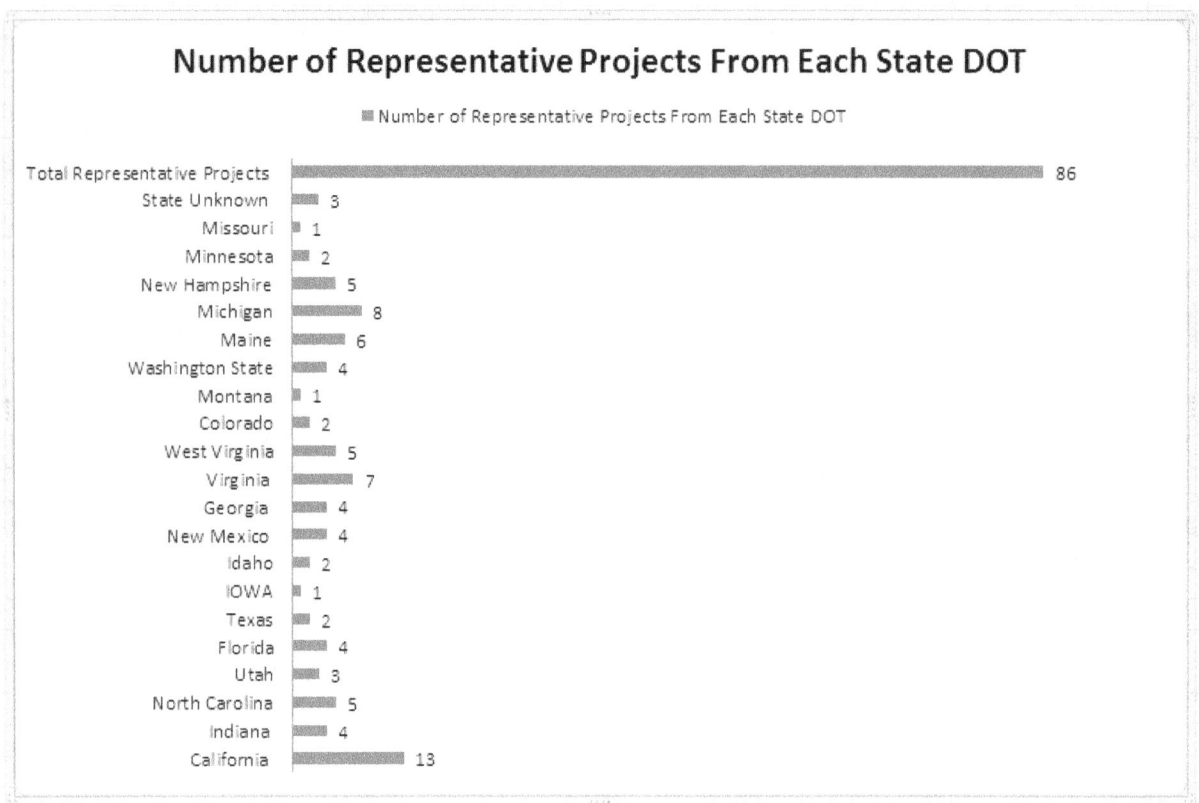

Figure 10. Number of Representative Projects Provided by the State DOTs

While it can be argued that most state DOTs are doing everything possible to address the issue of poor contract administration, unfortunately this belief was not reflected in the number of state DOT projects that participated, when compared to the number of projects executed by the state DOTs annually.

Appendix F indicates the proportion of responses received from respondents on all 40 contract administration practices. One of the key findings from the study was that on the average the practices in the questionnaire are applicable (excluding "Not Applicable" responses) to more than 84 percent the respondents, which confirms that the items do apply to most state DOTs. The finding indicates that all the items are in practice, and that the rate of practice in place ("YES" responses) range from 32 percent to 99 percent depending on the item.

Table 6 Contract Administration Practices and the Degree of Practice by the Sampled State DOT

Projects in the U.S.

Variables and Indicators	Practice was In Place	Practice was Not In Place	Practice was Reported as Not Applicable	State of Practice was Not Reported
Management Attitude Towards Contract Risks				
Organize Additional Partnering Sessions if Required	42%	29%	29%	0%
Provide Significant Contributors at Partnering Sessions	60%	3%	36%	0%
Provide Competent Superintendents	91%	8%	0%	1%
Provide Competent Owner Representatives	95%	3%	1%	0%
Provide Mutual Contracting Environment	92%	2%	5%	1%
Provide Periodic Evaluations	81%	17%	1%	0%
Pay Undisputed and Executed Item of Work Promptly	97%	2%	1%	0%
Contract Provisions For Mitigating Contract Risks				
Include Markup Provisions in the Contract	94%	1%	5%	0%
Include Time Related Overhead (TRO) Provisions in the Contract	33%	52%	8%	7%
Include Partnering Provisions in the Contract	42%	44%	14%	0%
Include Final Inspection and Closeout Provisions in the Contract	98%	2%	0%	0%
Include Provisions to Mitigate Potential Change Scenarios	99%	1%	0%	0%
Stability of Scope Definition				
Maintain Low Team Turnover	55%	13%	23%	9%
Maintain Dedicated Design and Engineering Team	86%	10%	3%	0%
Reduce the Degree of Scope Changes	77%	17%	5%	1%
Contract Administration Infrastructure				
Aggregate Direct Cost of Labor Properly	56%	17%	19%	8%
Support Changes with the Right Documentation	80%	17%	2%	0%
Aggregate Change Order Work and Original Contract Properly	66%	14%	9%	10%

Table 6 (Continued)

Variables and Indicators	Practice was In Place	Practice was Not In Place	Practice was Reported as Not Applicable	State of Practice was Not Reported
Resource Allocation Strategy				
Assign Resources Promptly When Changes Are Encountered	79%	13%	8%	0%
Assign Knowledgeable Superintendents with Authority to Resolve Disputes	73%	5%	22%	0%
Assign Contract Administration Team at the Onset of Project	87%	9%	1%	2%
Competency of Contract Administrators				
Convene Dispute Resolution Board (DRB) Members on Time	36%	9%	52%	2%
Comply with False Claims Awareness and Poster Provisions	55%	19%	12%	15%
Submit Change Order Based on Pre-established Provisions	50%	37%	12%	1%
Avoid Practices that Lead to Payment Withholding	52%	29%	16%	2%
Exclude Cost of Consultant and Legal Fee in Change Order Pricing	70%	8%	20%	2%
Comply with Certified Payroll Provisions	69%	21%	8%	2%
Comply with DBE Provisions	70%	15%	12%	3%
Document Daily Extra Work Sheets	63%	3%	34%	0%
Comply with Buy America Act Provisions	79%	7%	10%	3%
Comply with Applicable State, Federal and Local Laws	95%	1%	1%	2%
Use Measured Mile Method Where Applicable	20%	12%	63%	6%
Submit Update Schedules on Time	57%	33%	9%	1%
Follow TIA Preparation Requirements	28%	13%	51%	8%
Submit Baseline Schedule on Time	81%	12%	3%	3%
Transmit Notice Of Potential Claims on Time	57%	10%	30%	2%
Notify Owner of Changes on Time	72%	14%	12%	2%
Use WBS in Preparation of Project Schedules	59%	23%	15%	2%
Train Administrators in Partnering Practices	42%	12%	38%	8%
Use Visual Aids to Support Potential Claims	20%	40%	36%	5%

From Table 6 descriptive information can be reported on the degree of contract administration practices on state DOT projects. For detailed description of each practice as presented to respondents in the questionnaire, refer to appendices A, F and H.

It was found that only 42 percent of the respondents indicated that when applicable, additional partnering sessions were held to maintain partnering environment. However, 92 percent of respondents answered that us versus them attitudes were discouraged by providing a contracting environment of shared trust, equity, and commitment.

Regarding compliance with the requirement on false statement concerning highway projects, that ask if the notice on 18. U. S.C 1020 was posted on project regarding false claim, only 55 percent of the respondents had this practice in place. Ninety-nine percent of the respondents stated that applicable contract administration provisions (legal relations, changes, dispute resolutions, payments, progress schedule) were clearly outlined in the standard specifications and/or special provisions.

Regarding if the contractor assigned a competent contract administration team to the project at the start of project instead of waiting until changes are encountered, 87 percent of respondent answered yes.

Eighty-one percent of the respondents indicated that the baseline schedule was submitted by the due date. However, only 57 percent of the respondents reported that update schedules were submitted monthly, were current and were no more than one (1) month behind the due date.

Regarding time impact analysis, 28 percent of the respondents indicated that time impact analysis (TIA) submitted by the contractor meet specified TIA preparation requirements.

One practice that was least practiced by the respondents related to use of visual charting, graphing or other visual representation when communicating and supporting potential claims, and the rate of practice was only 20 percent.

Appendix G provides important information on rate of scope change, and average duration to address changes on state DOT projects. The sample indicated that 90 percent of the sampled projects experienced changes in scope at some point in time. This is an indication that there is a high probability that state DOT projects may most likely encounter some form of scope change.

The report also showed that the average variance resulting from change was about 5 percent, indicating that the original contract value was likely to change in value at this rate. It also showed that the average number of change orders encountered in a given year was thirteen.

Another important finding was that the average duration from discovery of change to the formal execution of a change order was two (2) months. This information should provide a baseline for general contractors to evaluate their ability to resolve changes quickly.

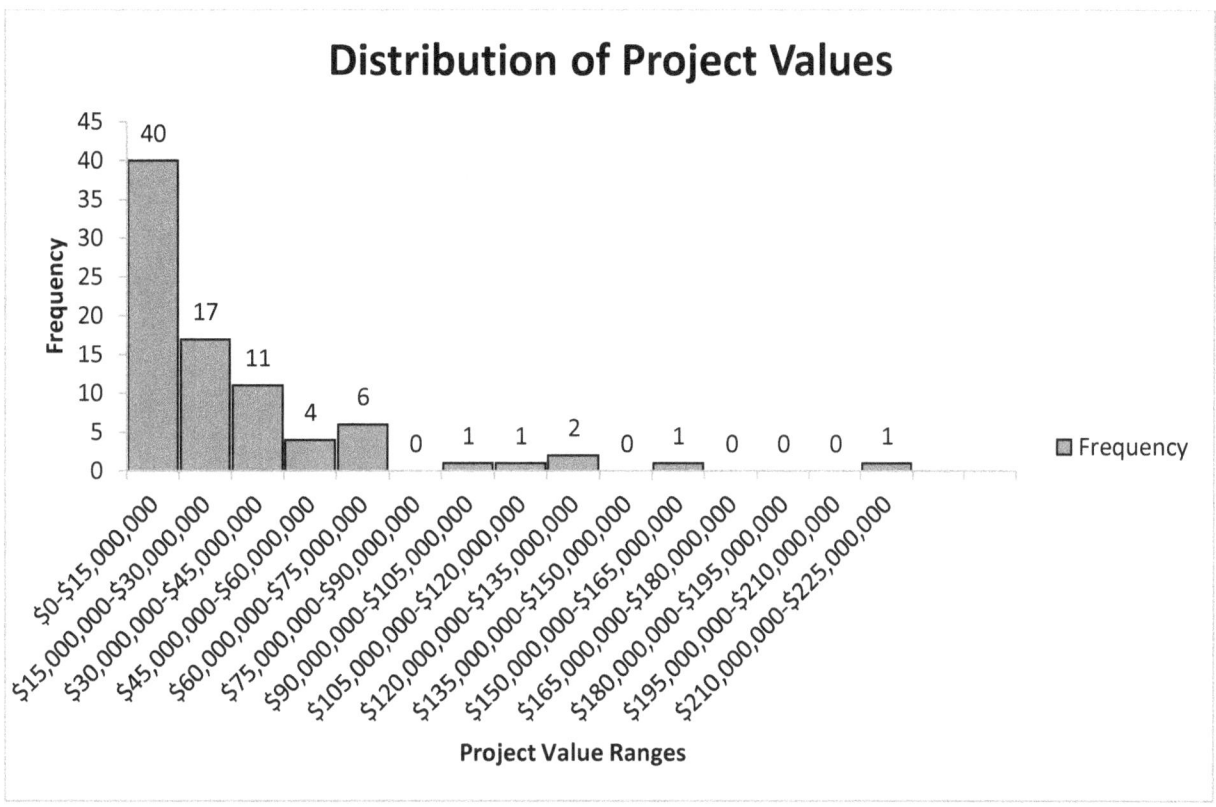

Figure 11. Project Values of Sampled Projects

The descriptive findings of sampled projects as shown in Figure 11 indicate that the majority of the projects sampled have project values less than $90M, which is a good representation of state DOT capital improvement projects for construction of roads and bridges.

From Descriptive Statistics to Inferential Statistics

The number of projects that participated in this study was 86, however not all of those projects were well-suited for inferential statistics. The samples were evaluated for missing data, and for practices that are not applicable to most state DOT projects. Three criteria were used in choosing samples and items included in statistical analyses. First, samples with missing data for the dependent variable were removed. Secondly, samples with more than 10 percent of missing data for the independent variables were removed. Finally, items that were not applicable to more than 30 percent of the respondents were removed. Based on this evaluation, the number of samples used for inferential statistics was reduced to 66 cases.

As shown in Appendix H, the instrument has 40 items and each item was measured on a "yes/no/not applicable scale and responses were coded 1 for "yes" and 0 for "no" and "not applicable".

CORRELATION ANALYSIS

This section presents the result of the correlation analysis. Using SPSS 19, the collected data was statistically analyzed using a correlation method to evaluate if the data indicated any relationship. The result of the statistical analysis using a correlation method allowed the study to make inference on whether a relationship does not exist.

Before the correlation analysis was conducted, data for the dependent variable was recoded to reflect the required analyses, and directly help to evaluate the study hypotheses. Contract administration performance was operationalized based on cycle time- from discovery of change order to execution of change order, which means that lower cycle time indicates higher contract administration performance level, and vice versa. The study hypotheses indicated that a positive correlation will be found within the population of interest, and the dependent variable was transformed accordingly. Appendix I shows how the operationalized data for the dependent variable was transformed by way of recoding.

Hypothesis No. 1

P1 – Ha: Management attitude towards contract risks is positively correlated to contract administration performance

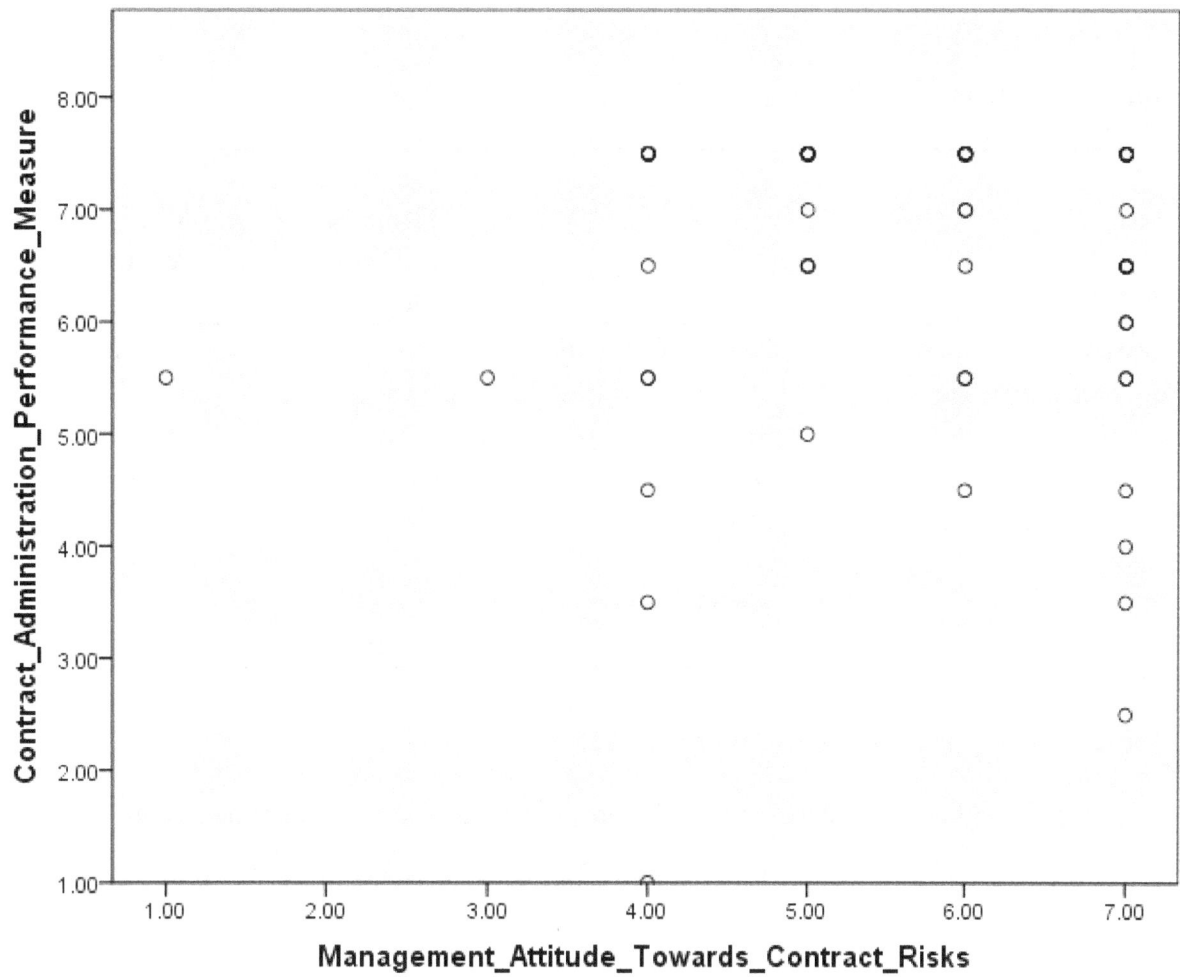

Figure 12. Management Attitude Towards Contract Risks and Contract Administration

Performance

Table 7 Descriptive Statistics for Hypothesis 1

	Mean	Std. Deviation	N
Contract Administration Performance Measure	6.4444	1.40882	63
Management Attitude Towards Contract Risks	5.5873	1.31535	63

Table 8 Correlation Result for Hypothesis 1

		Contract Administration Performance Measure	Management Attitude Towards Contract Risks
Contract Administration Performance Measure	Pearson Correlation	1.00	0.022
	Sig. (1-tailed)		0.431
	N	63.00	63.00
Management Attitude Towards Contract Risks	Pearson Correlation	0.022	1.00
	Sig. (1-tailed)	0.431	
	N	63.00	63.00

Discussion on Hypothesis No. 1, Figure 12, Table 7 and 8

This hypothesis aimed to test the relationship between contract administration performance and management attitude towards contract risks as they relate to actions by management to stay involved and participate in project outcome. It would be expected that contract administration performance would be positively improved when management gets involved and take necessary actions to support the projects.

No significant correlation was found, and this was surprising and in contrast to the hypothesized expected result.

Based on the collected data, the correlation analysis did not support the expected result and the magnitude of the relationship was not significant. It can be inferred that the collected data was not large enough to capture all the possible scenarios and create a pattern that would explain and support this fundamental relationship. Lack of significant correlation may be due to wrong assumption of linear relationship when in fact a nonlinear relationship existed. Another explanation includes the lack of quality indicators at capturing the underlying variables, or that

the influencing indicators were omitted. Also lack of significant correlation may be due to

measurement error related to clarity of the questions, or maybe there is no relationship.

Hypothesis No. 2

P2 – Ha: Contract provisions for mitigating contract risks are positively correlated to contract

administration performance

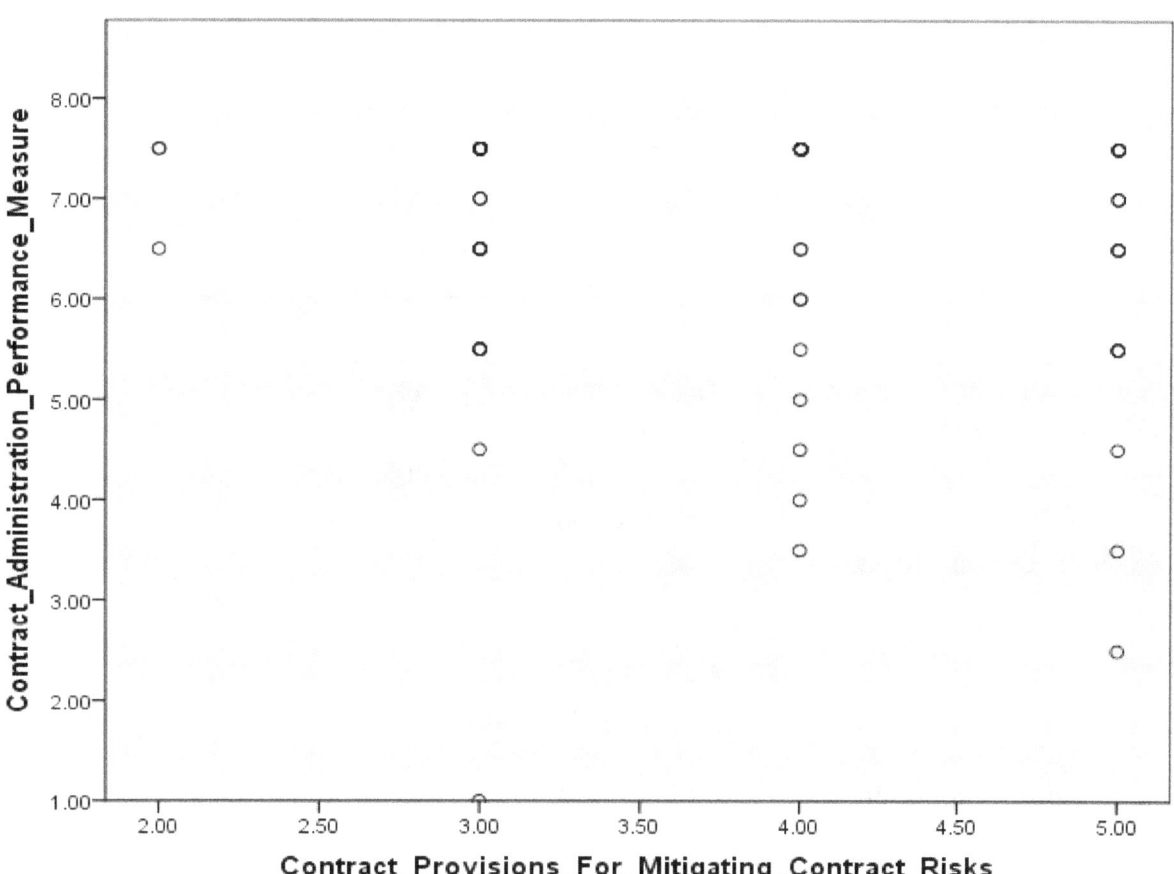

Figure 13. Contract Provisions for Mitigating Contract Risk and Contract Administration Performance

Table 9 Descriptive Statistics for Hypothesis 2

	Mean	Std. Deviation	N
Contract Administration Performance Measure	6.4444	1.40882	63
Contract Provisions For Mitigating Contract Risks	3.7302	0.88366	63

Table 10 Correlation Result for Hypothesis 2

		Contract Administration Performance Measure	Contract Provisions For Mitigating Contract Risks
Contract Administration Performance Measure	Pearson Correlation	1.00	-0.187
	Sig. (1-tailed)		0.071
	N	63.00	63.00
Contract Provisions For Mitigating Contract Risks	Pearson Correlation	-0.187	1.00
	Sig. (1-tailed)	0.071	
	N	63.00	63.00

Discussion on Hypothesis No. 2, Figure 13, Table 9 and 10

This hypothesis aimed to test the relationship between contract administration performance and contract provisions for mitigating contract risks as they relate to actions by management to include the right contract provisions in the contract as a way to mitigate contractual risks. It would be expected that contract administration performance will be positively improved when the right contract provisions are in place to mitigate contract risks.

No significant correlation was found to exist between contract administration performance and contract provisions for mitigating contract risks. This was surprising and in contrast to the hypothesized expected result.

Based on the collected data, the correlation analysis did not support the expected result and the magnitude of the relationship was not significant. It can be inferred that the collected data was not large enough to capture all the possible scenarios and create a pattern that would explain and support this fundamental relationship. Lack of significant correlation may be due to wrong assumption of linear relationship when in fact a nonlinear relationship existed. Another explanation includes the lack of quality indicators at capturing the underlying variables, or that the influencing indicators were omitted. Also lack of significant correlation may be due to measurement error related to clarity of the questions, or maybe there is no relationship.

Hypothesis No. 3

P3 – Ha: Stability of scope definition is positively correlated to contract administration performance

118

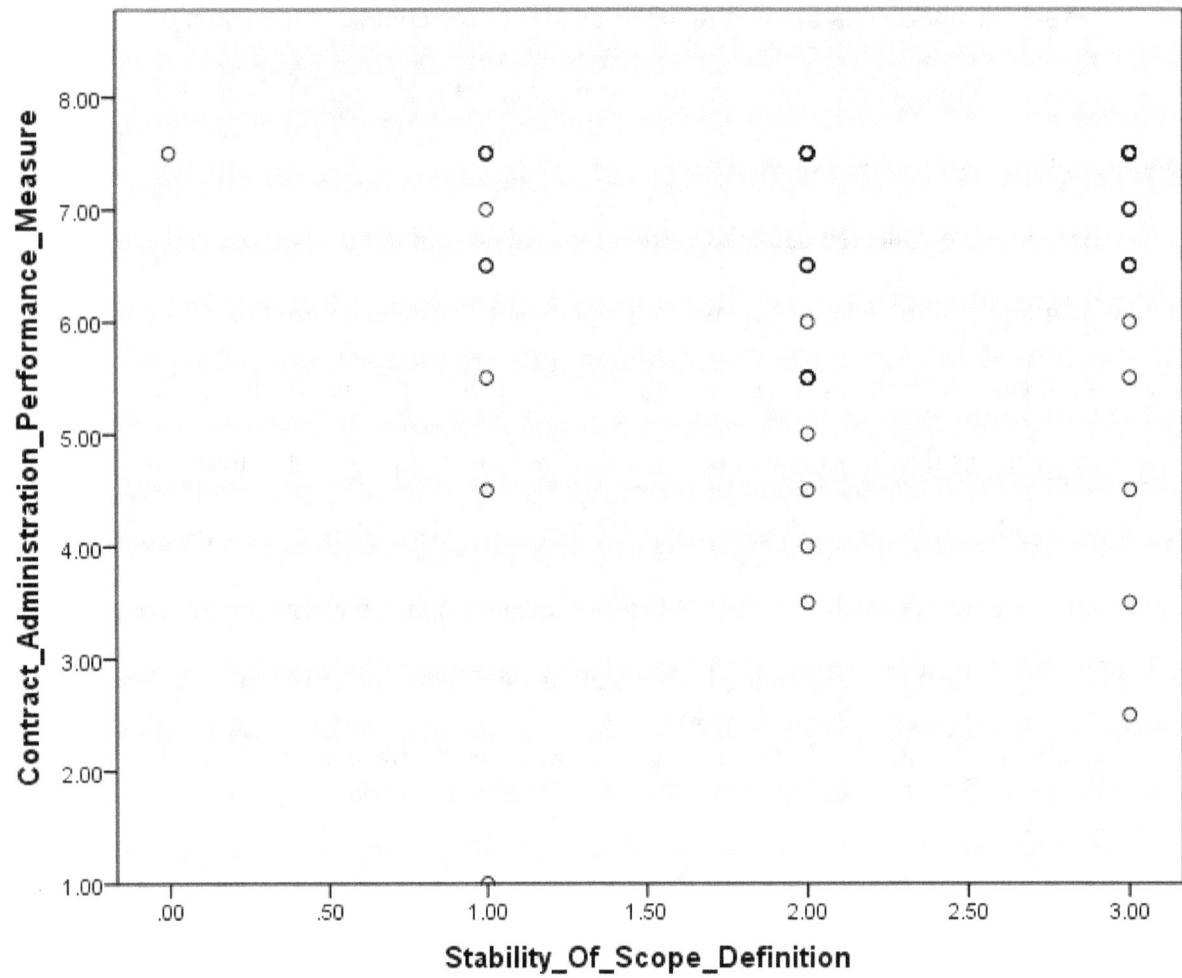

Figure 14. Stability of Scope Definition and Contract Administration Performance

Table 11 Descriptive Statistics for Hypothesis 3

	Mean	Std. Deviation	N
Contract Administration Performance Measure	6.4444	1.40882	63
Stability Of Scope Definition	2.2698	0.76636	63

Table 12 Correlation Result for Hypothesis 3

		Contract Administration Performance Measure	Stability Of Scope Definition
Contract Administration Performance Measure	Pearson Correlation	1.00	0.149
	Sig. (1-tailed)		0.123
	N	63.00	63.00
Stability Of Scope Definition	Pearson Correlation	0.149	1.00
	Sig. (1-tailed)	0.123	
	N	63.00	63.00

Discussion on Hypothesis No. 3, Figure 14, Table 11 and 12

This hypothesis aimed to test the relationship between contract administration performance and stability of scope definition and requirements as they relate to actions by management to stabilize scope and requirements so that they can be addressed effectively. It would be expected that contract administration performance will be positively improved when management take actions to stabilize scope definition and requirements.

No significant correlation was found to exist between contract administration performance and stability of scope definition and requirements. This was surprising and in contrast to the hypothesized expected result.

Based on the collected data, the correlation analysis did not support the expected result and the magnitude of the relationship was not significant. It can be inferred that the collected data was not large enough to capture all the possible scenarios and create a pattern that would explain and support this fundamental relationship. Lack of significant correlation may be due to wrong assumption of linear relationship when in fact a nonlinear relationship existed. Another

120

explanation includes the lack of quality indicators at capturing the underlying variables, or that

the influencing indicators were omitted. Also lack of significant correlation may be due to

measurement error related to clarity of the questions, or maybe there is no relationship.

Hypothesis No. 4

P4 – Ha: Contract administration infrastructure is positively correlated to contract administration performance

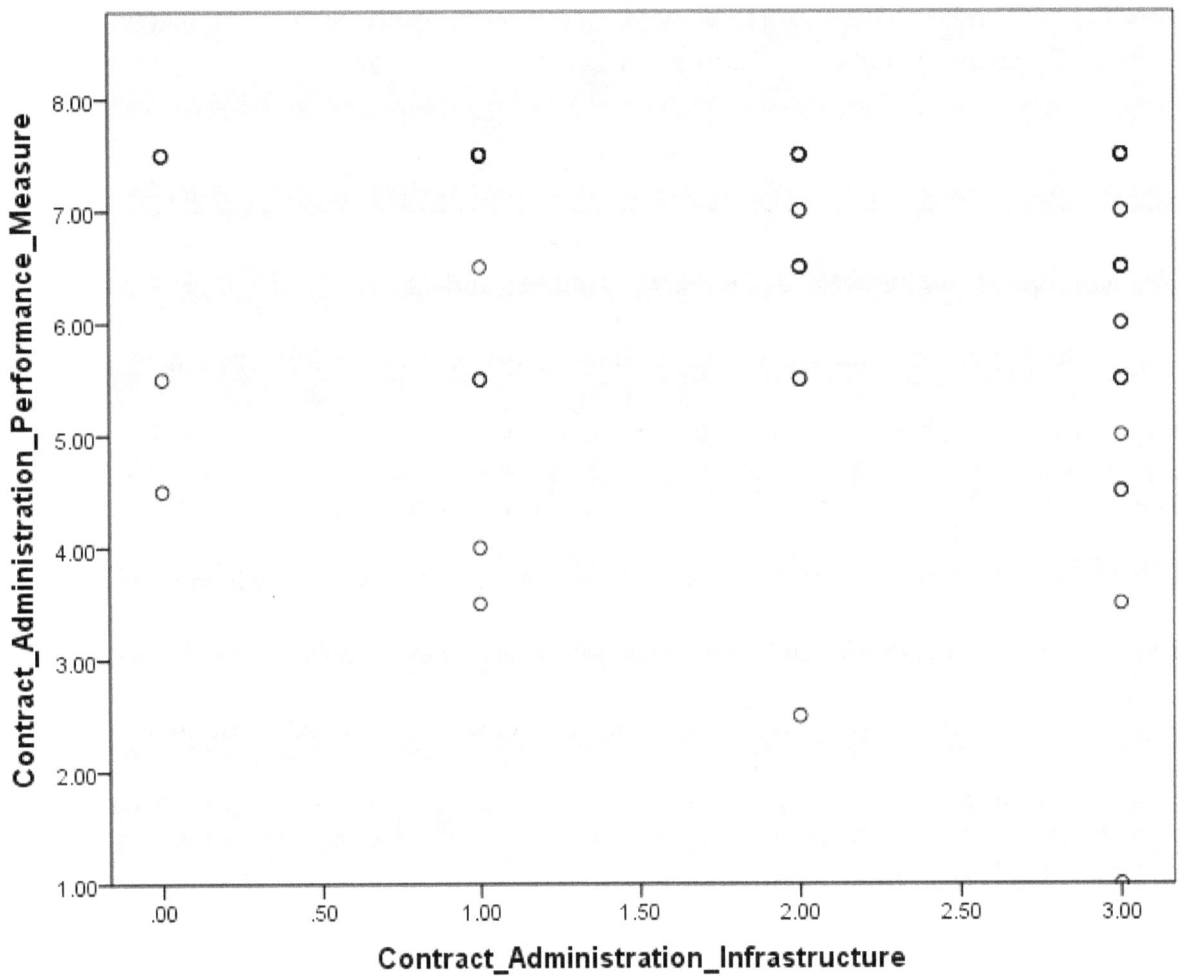

Figure 15. Contract Administration Infrastructure and Contract Administration Performance

Table 13 Descriptive Statistics for Hypothesis 4

	Mean	Std. Deviation	N
Contract Administration Performance Measure	6.4444	1.40882	63
Contract Administration Infrastructure	2.0635	0.99795	63

Table 14 Correlation Result for Hypothesis 4

		Contract Administration Performance Measure	Contract Administration Infrastructure
Contract Administration Performance Measure	Pearson Correlation	1.00	-0.095
	Sig. (1-tailed)		0.230
	N	63.00	63.00
Contract Administration Infrastructure	Pearson Correlation	-0.095	1.00
	Sig. (1-tailed)	0.230	
	N	63.00	63.00

Discussion on Hypothesis No. 4, Figure 15, Table 13 and 14

This hypothesis aimed to test the relationship between contract administration performance and contract administration infrastructure as they relate to actions by management to make sure that the right technologies are in place for document control, cost control and schedule control. It would be expected that contract administration performance will be positively improved when the right technologies are in place to manage and control costs, time and documents.

No significant correlation was found to exist between contract administration performance and contract administration infrastructure. This was surprising and in contrast to the hypothesized expected result.

Based on the collected data, the correlation analysis did not support the expected result and the magnitude of the relationship was not significant. It can be inferred that the collected data was not large enough to capture all the possible scenarios and create a pattern that would explain and support this fundamental relationship. Lack of significant correlation may be due to wrong assumption of linear relationship when in fact a nonlinear relationship existed. Another explanation includes the lack of quality indicators at capturing the underlying variables, or that the influencing indicators were omitted. Also lack of significant correlation may be due to measurement error related to clarity of the questions, or maybe there is no relationship.

Hypothesis No. 5

P5 – Ha: Resource allocation strategy is positively correlated to contract administration performance

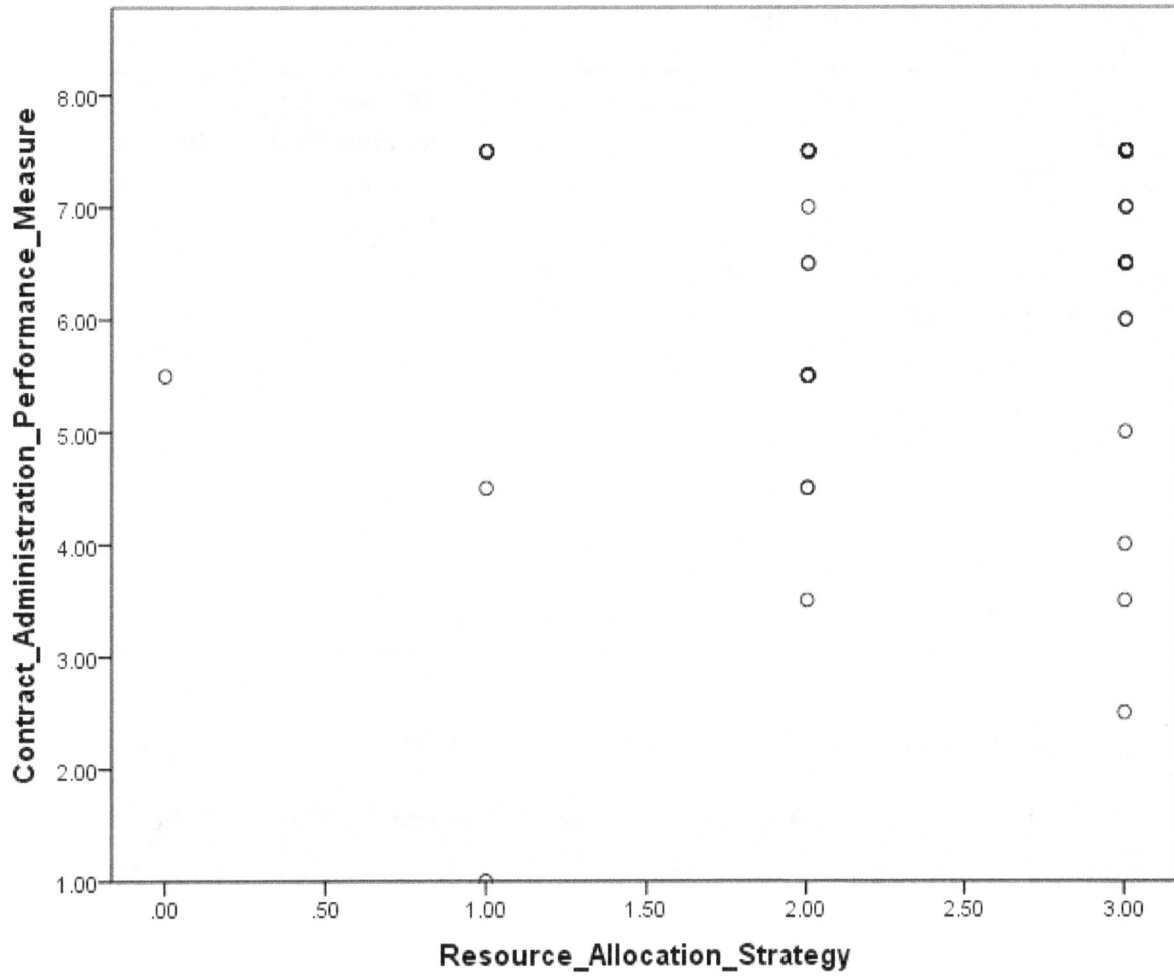

Figure 16. Resource Allocation Strategy and Contract Administration Performance

Table 15 Descriptive Statistics for Hypothesis 5

	Mean	Std. Deviation	N
Contract Administration Performance Measure	6.4444	1.40882	63
Resource Allocation Strategy	2.4444	0.73568	63

Table 16 Correlation Result for Hypothesis 5

		Contract Administration Performance Measure	Resource Allocation Strategy
Contract Administration Performance Measure	Pearson Correlation	1.00	0.227[*]
	Sig. (1-tailed)		0.037
	N	63.00	63.00
Resource Allocation Strategy	Pearson Correlation	0.227[*]	1.00
	Sig. (1-tailed)	0.037	
	N	63.00	63.00
*. Correlation is significant at the 0.05 level (1-tailed).			

Discussion on Hypothesis No. 5, Figure 16, Table 15 and 16

This hypothesis aimed to test the relationship between contract administration performance and resource allocation strategy as they relate to actions by management to make sure that the right resources are assigned at the right time. It would be expected that contract administration performance will be positively improved when management take actions to allocate the right resources at the right time.

The correlational analysis was significant and positive, which was in agreement with the hypothesized expected result, indicating that performance will improve when management takes the right steps to allocate the right resources at the right time.

Based on the collected data, the correlation analysis supported the expected result and the magnitude of the relationship was significant. It can be inferred that the collected data captured the possible scenarios and pattern that would explain and support this fundamental relationship. The significant correlation supported the assumption of linear relationship. This result also indicated the presence of quality indicators for capturing the underlying variables. Also significant correlation indicated that the sub-scale was valid.

The study finding showed only hypothesis 5 as significant, even though the sample size was pointed out as one of the reasons why the other hypotheses were not significant and a discussion may be appropriate here on why hypothesis 5 was significant even with a small sample size. Hypothesis 5 relates to resource allocation strategy, and it has indicators that include: assignment of competent administrators at the start of the project; assignment of competent administrators to manage changes as soon as they are encountered; and assignment of competent administrators during resolution of disputes. The study finding supports the understanding that human resources are the most important asset when the right people, with the right information are assigned at the right time to the right duty. Unlike the other factors, practices that operationalized resource allocation strategy showed a high rate of practice above 72 percent, and the difference between their rates of practice was small. This result might indicate some consistency in practice which could account for the significant finding.

Hypothesis No. 6

P6 – Ha: Contract administrators' competency is positively correlated to contract administration performance

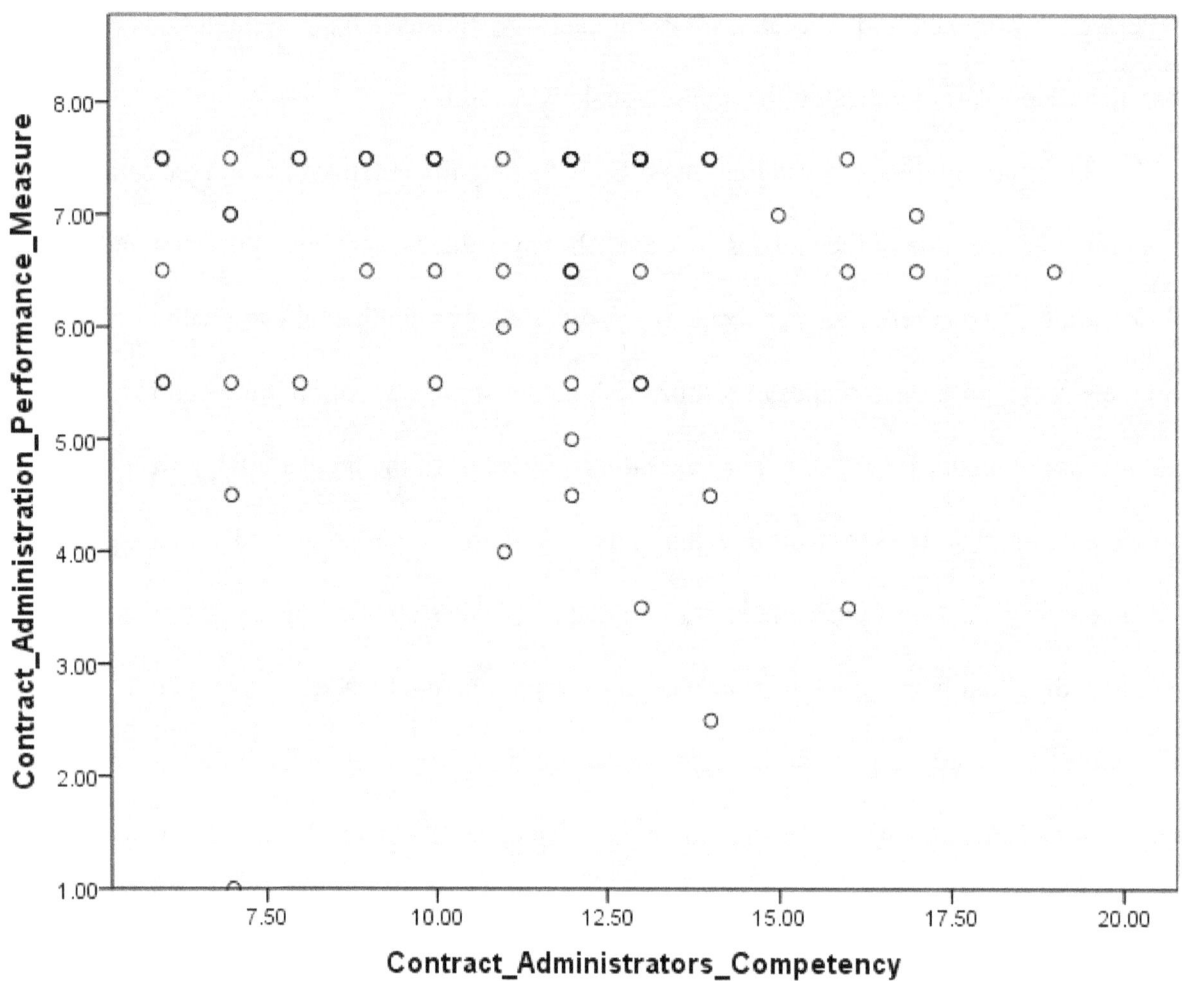

Figure 17. Contract Administrators' Competency and Contract Administration Performance

Table 17 Descriptive Statistics for Hypothesis 6

	Mean	Std. Deviation	N
Contract Administration Performance Measure	6.4444	1.40882	63
Contract Administrators Competency	11.1905	3.14621	63

Table 18 Correlation Result for Hypothesis 6

		Contract Administration Performance Measure	Contract Administrators Competency
Contract Administration Performance Measure	Pearson Correlation	1.00	0.002
	Sig. (1-tailed)		0.492
	N	63.00	63.00
Contract Administrators Competency	Pearson Correlation	0.002	1.00
	Sig. (1-tailed)	0.492	
	N	63.00	63.00

Discussion on Hypothesis No. 6, Figure 17, Table 17 and 18

This hypothesis aimed to test the relationship between contract administration performance and competency of contract administrators as they relate to actions by contract administrators to effectively manage contracts. It would be expected that contract administration performance will be positively improved when the contract administrators are competent.

No significant correlation was found to exist between contract administration and competency of contract administration. This was surprising and in contrast to the hypothesized expected result.

Based on the collected data, the correlation analysis did not support the expected result and the magnitude of the relationship was not significant. It can be inferred that the collected data was not large enough to capture all the possible scenarios and create a pattern that would explain and support this fundamental relationship. Lack of significant correlation may be due to wrong assumption of linear relationship when in fact a nonlinear relationship existed. Another explanation includes the lack of quality indicators at capturing the underlying variables, or that the influencing indicators were omitted. Also lack of significant correlation may be due to measurement error related to clarity of the questions, or maybe there is no relationship.

MULTIPLE REGRESSION ANALYSIS

The second objective of this study was to develop a predictive model for predicting and controlling contract administration performance of general contractors on federal and state DOT projects in the U.S. However, in order to develop a model, there should be a significant correlation between the dependent and independent variables. The previous section on correlation analysis shows that based on the data collected, a significant correlation was found only between contract administration performance and resource allocation strategy. However, an investigation using regression analysis revealed that the collected data were not suitable for development of a predictive model.

CHAPTER 5

CONCLUSION: APPLYING THE STUDY FINDINGS TO IMPROVE PERFORMANCE

This section is the end product or output of the research study. It aims to report on how well the research objectives were met, how well the hypotheses were aligned with what is going on, practical application of the research findings, issues that may have limited or hindered the study from meeting the research objectives, and finally, recommendation for further study.

RESEARCH OBJECTIVES

The overall research question addressed in this study was: Is there a relationship between contract administration practices and contract administration performance of general contractors on federal and state DOT projects in the U.S.? To answer the question posed above, the study sought to accomplish the following objectives-

1. Identify contract administration practices that are associated with general contractors' ability to meet contract administration objectives consistently on federal and state DOT projects in the U.S. and

2. Examine if an association exists between the dependent and independent variables, and if so, to design a predictive model that can be used to predict and control general contractor's ability to meet contract administration objectives consistently on federal and state DOT projects in the U.S.

The study met the first objective and identified that there was a significant correlation between contract administration performance and resource allocation strategy; however, the relationships were not statistically significant for the other variables.

The second objective was dependent on finding statistically significant relationship between the dependent and independent variable. The study was unable to design the predictive model because the regression test indicated that the collected data were not suitable for the development of a predictive model.

RESEARCH HYPOTHESES

The study hypotheses were based on well-defined research questions and were simply stated to reflect one dependent variable to one independent variable. Also they were specifically stated without ambiguity on the variables and the population of interest. With respect to a population of interest consisting of general contractors working on federal and state DOT projects, the tentative theories of this study included the following:

P1 – Ha: Management attitude towards contract risks is positively correlated to contract administration performance

P2 – Ha: Contract provisions for mitigating contract risks are positively correlated to contract administration performance

P3 – Ha: Stability of scope definition is positively correlated to contract administration performance

P4 – Ha: Contract administration infrastructure is positively correlated to contract administration performance

P5 – Ha: Resource allocation strategy is positively correlated to contract administration performance

P6 – Ha: Contract administrators' competency is positively correlated to contract administration performance

Based on data collected for this study, the correlational analysis only supported hypothesis No. 5. However the correlation analysis could not confirm hypotheses No. 1,2,3,4 and 6.

RESEARCH CONTRIBUTIONS AND USAGE

The first key finding from this study was that a significant correlation existed between contract administration performance and resource allocation strategy. However, the investigation showed that the result of the regression analysis was not suitable for development of a predictive model. Since a significant correlation was found between contract administration performance and resource allocation strategy, state DOT projects can use this as a tool to evaluate how well they are doing.

The second key finding from the study was that the average cycle time from discovery of change to formal execution of change order was two (2) months, and this can be used on state DOT projects as the baseline for evaluating performance level.

The third key finding from the study was that on the average, the practices in the questionnaire were applicable to more than 84 percent of the respondents, which confirms that the items do apply to most state DOTs, and can be streamlined by each state DOT for performance evaluation.

State DOTs should use the cycle time of two (2) months as a baseline to assess how quickly their projects are addressing and resolving change orders. On the other hand, using the questionnaire, each state DOT should identify what contract administration practices apply to their projects. The agencies could then use the list as a checklist to evaluate their contract administration performance level on each project.

LIMITATIONS

The main limitation encountered by this study was the lack of access to larger number of samples. The sample size may not have been large enough to capture the possible scenarios and expose the patterns that underlie this phenomenon. Poor participation by state DOT in this study may in fact have hindered the study.

RECOMMENDATIONS

Due to the low participation in this study, I would recommend that a committee of state DOT contract administrators be set up to replicate this study. Each state DOT would coordinate and help gather required data, and make sure that majority of the projects executed by their agency are involved in the study. Also, the state DOTs are encouraged to replicate this study by conducting an in depth analysis on a state-by-state basis.

Secondly, since the state DOT projects are publicly funded by the taxpayers and the agencies seek to be transparent, I recommend that the state DOTs make basic contract data available to the public and to researchers specifically. For example, data on original contract amount, current contract amount, average duration to resolve change orders, complexity of change order, type of scope change (E.W-extra work or A.C-adjustment of compensation), etc., should be available to the public. This is important because lack of access to data is a hindrance to researchers seeking to understand and find a solution to the problem of contract administration performance.

I recommend that state DOTs conduct monthly performance evaluations using the items identified in the research questionnaire as a starting point, with option to add, remove or modify the items to fit their situation.

REFERENCES

Adams, K. F. (2008). Construction contract risk management: A study of practices in the United Kingdom. *Cost Engineering Journal*, *50*(1), 22-33

Avery, S. (2004). Contract management is the next step in smart supply strategy. *Purchasing Magazine*, *133*(12), 60-64

Bauer. W. (2005). Going the extra mile. *Cost Engineering Journal,* *47*(11), 17-21

Belassi, W., & Tukel, O. (1996). A new framework for determining critical success/failure factors in projects. *International Journal of Project Management*, *14*(3), 141-151

Bertalanffy, L. V. (1972). The history and status of general systems theory. *Academy of Management Journal*, *15*(4), 407-426,

Boulding, K. E. (1956). "General systems theory - the skeleton of science," *Management Science*, 2: 197-208

Bourdeau, L. (1999). Sustainable development and the future of construction: A comparison of visions from various countries. *Building Research & Information,* *27*(6), 354–366

Brown, M. (1996). *Keeping score: Using the right metrics to drive world-class performance.* New York: Quality Resources.

Bryman, A. & Cramer, D. (2011). *Quantitative data analysis with IBM SPSS 17, 18 & 19 : A guide for social scientists*. New York: Routledge.

Capra, F. (2006). From the parts to the whole: Systems thinking in ecology and education. Retrieved from http://www.hainescentre.com/pdfs/parts_to_whole.pdf

Cardullo, M. (1996). *Introduction to managing technology*. Forest Grove: Research Studies

Press.

Chan, A. P. C., Scott, D., & Chan, A. P. L. (2004). Factors affecting the success of a construction project. *Journal of Construction Engineering and Management*, *130*(1), 153-155.

Cheng, E. W. L., & Li, H. (2002). Construction partnering process and associated critical success factors: Quantitative investigation. *Journal of Management in Engineering*, *18*(4), 194-202

Cheung, S. O., & Suen, H. C. H. (2002). A multi-attribute utility model for dispute resolution strategy selection. *Construction Management and Economics, 20*(7), 557-68.

Cheung, S. O., Tam, M. C., & Harris, C. F. (2000). Project dispute resolution satisfaction classification through neural network. *Journal of Management in Engineering*, *16*(1), 70-79.

Charoenngam, C., Coquinco S. T. & Hadikusumo, B. H. W. (2003). Web-based application for managing change orders in construction projects. *Construction Innovation, 3*, 197–215.

Cox. K. R. (1997). Managing change orders and claims. *Journal of Management in Engineering*, Jan/Feb 1997. p24-29

Criss, M. R. (2006). Contract compliance for government contractors. *Contract Management, 46*(12), 14-20

Davis, M., Aquilano, N., & Chase, R. (2002). *Fundamentals of operations management*. Boston: Irwin McGraw-Hill.

Dawson. B. (2007). Bertalanffy Revisited: Operationalizing A General Systems Theory Based Business Model Through General Systems Thinking, Modeling, And Practice. Proceedings of the 51st Annual Meeting of the ISSS, Papers: 51st Annual Meeting Tokyo, Japan, 2007. ISSN 1999-6918

Delano, J. K. M. (1998). Identifying factors that contribute to program success. *Acquisition Review Quaterly*. Winter 1998

DeVellis, R. (2003). *Scale development*. Thousand Oaks: Sage.

Dysert, L. (2007). Is estimate accuracy and oxymoron? *Cost Engineering Journal*, *49*(1), 32-36

El-adaway, H. I., & Kandil, A. A. (2009). Contractors' claim insurance: A risk retention approach. *Journal of Construction Engineering and Management*, *135*(9), 819-825.

Essens, P., Vogelaar, A., Mylle, J., Blendell, C., Paris, C., Halpin, S., & Baranski, J. (2005). *Military command team effectiveness: Model and instrument for assessment and improvement*. NATA RTO HFM-087/RT6-023

Few, S. (2009). *Now you see it: simple visualization techniques for quantitative analysis*. Oakland, Calif: Analytics Press.

Fernández-Solís, J. (2009). How the construction industry does differ from manufacturing. 2009 *Associated School of Construction (ASC) Conference Proceedings* at University of Florida.

Floyd, L. A. (2004). Application of appropriate project controls tools for contract type. *Cost Engineering Journal, 46*(2), 25-30

Fínez, J. F. (2008). Three step methodology to measure an individual's personal competency for entrepreneurship towards a "particular" business idea. *Journal of Technology Management &. Innovation, 3*(1), 99-107.

Fisk, R. E., & Reynolds, D. W. (2010). *Construction project administration*. Englewood Cliffs: Prentice Hall.

Fullerton, R. (2005). Searching for balance in conflict management: The contractor's perspective. *Dispute Resolution Journal, 60*(1), 48-61.

Garrett, A. G. (2010). Contract administration, part 1: People, processes and best practices. *Contract Management*. February 2010. p52-66

Garrett, A. G. (2007). Post-award contract administration: Lessons learned and best practices. *Contract Management*. July 2007, p36-44

Georgy, M. E., Chang, L., & Zhang, L. (2005). Predition of engineering performance: A neurofuzzy approach. *Journal of Construction Engineering and Management, 131*(5), 548-557.

Glagola, R. C., & Sheedy, M. W. (2002). Partnering on defense projects. *Journal of Construction Engineering and Management, 128*(2), 127-138.

Goetz, J. C., & Gibson, G. E. (2009). Construction litigation, U.S. General Service Administration, 1980-2004. *Journal of Legal Affairs and Dispute Resolution in Engineering and Construction, 1*(1), 40-46.

Gray, C., & Larson, E., (2006). *Project management*. Boston: McGraw-Hill/Irwin

Gransberg, D. D., & Villarreal Builrago, M. E. (2002). Construction project performance metrics. *AACE International Transaction*. Morgantown, WV: AACE International CSC.02 p1-5

Griffith, F. A. (2006). Scheduling practices and project success. *Cost Engineering Journal*, *48*(9), 24-30

Hanna, S. A., Camlic, R., Peterson, A. P., & Nordheim, V. E. (2002). Quantitative definition of projects impacted by change orders. Journal of Construction Engineering and Management, *128*(1), p57-64

Hassanein, G. A. A., & Afify, F. M. H. (2007). Contractors' perceptions of construction risk: A case study of power station projects in Egypt. Cost Engineering Journal, *49*(5), 25-34.

Hassanein, G. A. A., & Nemr, E. W. (2009). Change order claims in the Egyptian industrial construction sector: Causes and cost/time overruns. *Cost Engineering Journal*, *51*(11), 21-29.

Havers, J., O'brien, J., & Stubbs, F. (1996). *Standard handbook of heavy construction*. New York: McGraw-Hill

Huselid, M. (1995). The impact of human resource management practices on turnover, productivity, and corporate financial performance. *Academy of Management Journal*, *38*(3), 635-872.

Ibbs, C. W., & Ashley, D. B. (1987). Impact of various construction contract clauses. *Journal of Construction Engineering and Management, 113*(3), 501-521.

International Standards Organization – (ISO) DIS 12006-2. (2001) Organization of information about construction works. Part 2 Framework for classification of information. ISO 2001

International Technology Education Association-ITEA (1996). *Technology for all Americans: A rational and structure for the study of technology*. ITEA, Reston. VA.

139

Jaselskis, E. J., & Ashley, D. B. (1991). Optimal allocation of project management resources for

achieving success. *Journal of Construction Engineering and Management, 117*(2), 321-

340.

Kaliprasad, M. (2006a). The human factor I: Attracting, retaining, and motivating capable

people. *Cost Engineering Journal, 48*(6), 20-26.

Kaliprasad, M. (2006b). The human factor II: Creating a high performance culture in an

organization. *Cost Engineering Journal, 48*(6), 27-34.

Kaliprasad, M. (2006c). Proactive risk management. *Cost Engineering Journal, 48*(12), 26-36.

Katz, G. I. (n.d). Risk management begins with your contract. Retrieved from

http://www.katzandstone.com/pdf/rmbeginswithyourcontract.pdf

Kazi, A., (2004). *Knowledge management in the construction industry*. City: IGI Global.

Khalil, T. (2000). *Management of technology – The key to competitiveness and wealth creation.*

New York: McGraw-Hill Companies.

Khan, A. (2006). Project scope management. *Cost Engineering Journal, 48*(6), 12-16.

Korde, T., Li, M., & Russell, A. (2005). State-of-the-art review of construction performance

models and factors. Proceedings, 2005 *ASCE Construction Research Congress, San*

Diego, USA.

Lewis, J. (2000). *The project manager's desk reference*. New York: McGraw-Hill.

Levin, P. (1998). *Construction contract claims, changes, & dispute resolution*. New York: ASCE

Li, M., Korde, T. & Russell, A. (2005). Explaining construction performance using causal

models. *Proceedings, CD Rom 6th Construction Specialty Conference, Canadian Society*

of Civil Engineers, June 2-4, 2005, Toronto, Canada, 12 pages, refereed

Li, G. W. (2005a). Construction baseline schedule review and submittal timeframe. *Cost Engineering Journal, 47*(2), 28-36.

Li, G. W. (2005b). A case study: Schedule forecast verification with productivity assessment. *Cost Engineering Journal, 47*(11), 22-30.

Ling, F. Y. Y., Low, S. P., Wang, S., & Egbelakin, T. (2008). Models for predicting performance in China using project management practices adopted by foreign AEC firms. *Journal of Construction Engineering and Management. 134*(12), 983-990.

Lozon, J., & Jergeas. G. (2008). The use and impact of value improving practices and best practices. *Cost Engineering Journal, 50*(6), 26-32.

Markert, L., & Backer, P. (2002). *Contemporary technology*: *Innovations, issues, and perspectives*. City: Goodheart-Wilcox Publisher.

Mehta, M. P., (2008). Effective implementation of a project control system. *Cost Engineering Journal, 50*(1), 34-37.

Miles, J., & Shevlin, M. (2001). *Applying regression & correlation*. Thousand Oaks: Sage Publications

Michel, H. (1998). The next 25 years: The future of the construction industry. *Journal of Management Engineering, 14*(5), 26-31.

Mohan, B.S., & Al-Gahtani, S. K. (2006). Current delay analysis techniques and improvements. *Cost Engineering Journal, 48*(9), 12-21.

Mohr, L. (1990). *Understanding significance testing*. Thousand Oaks: Sage Publications.

Molly, K. K. (2007). Six steps for successful change management. *Cost Engineering Journal, 47*(4), 12-19.

Nael G. B. (2003). Contract or cooperation? Insights from the Middle East. *A paper given to a conference organized by the Center of Construction Law* at King's College London, 11th September 2003.

Nalewaik, A. A. (2011). Control and audit devices for claims management. *Cost Engineering Journal, 53*(8), 18-22.

National Association of Surety Bond Producers. (2009). Why do contractors fail. Retrieved from http://www.sio.org/PDF/WhyFail.pdf

Norfleet, A. D. (2005). Loss of learning in disruption claims. *Cost Engineering Journal, 47*(11), 10-14.

Ofori-Boadu, A.N., Okere, G. & Kim, C (2010). BIM: Implementation strategies and future implications. *International Journal of Project Planning and Finance, 1*(1), 102-128.

Office of Federal Procurement Policy (1994). A guide to best practices for contract administration. Retrieved from https://www.acquisition.gov/bestpractices/bestpcont.html

Okere, O. G. (2010). The elusive goal of consistently meeting contract administration objectives: A conceptual report. *Cost Engineering Journal, 52*(4), 24-29.

Project Management Institute - PMI. (2002). *Project Manager Competency Development Framework* (PMCD). Project Management Institute.

Park, W., & Chapin, W. (1992). *Construction bidding*. New York: Wiley.

Perkins, P. (2008). The contractor performance evaluation system revisited. Federal Construction Project Manager's Bulletin. Prepared by Construction Contract Specialists. V3, No5

Portilla, A. (2010). Productivity loss for sewer pipe installation. *Cost Engineering Journal, 52*(4), 11-16.

Project on Government Oversight (POGO). (n.d). *Federal contract misconduct*. Retrieved from
http://contractormisconduct.org/

Quinn, F. (2005). The Power of procurement. *Supply Chain Management Review, 9*(9), 6-8.

Rad. F. P. (2003). Project success attributes. *Journal of Cost Engineering, 45*(4), 23-29.

Rad, F. P., & Cioffi, F. D. (2004). Work and resource breakdown structures for formalized
bottom-up estimating. *Cost Engineering Journal, 46*(2), 31-37.

Ren, Z., Anumba, C J., & Ugwu, O. (2000). Towards a multi-agent system for construction
claims negotiation. In: Akintoye, A (Ed.), *16th Annual ARCOM Conference*, 6-8
September 2000, Glasgow Caledonian University. Association of Researchers in
Construction Management, Vol. 1, 385-93.

Rummler, G., Ramias, A., & Rummler, R. (2009). *White space revisited: Creating value through
process*. San Francisco: Jossey-Bass.

Rummler, G., & Brache, A. (1995). *Improving performance*. San Francisco: Jossey-Bass
Publishers.

Rwelamila, D. P., Lobelo, L., & Kupakuwana, S. P. (2004). Insolvencies among civil
engineering enterprises in South Africa: A time to reflect. *Cost Engineering Journal,
46*(7), 12-14 &31-37.

Sanders, M. (2004). Systematic contract change review. *AACE International Transactions*

Sandlin, L. S., Sapple, J. R., & Gautreaux, R. M. (2004). Phased root cause analysis: A
distinctive view on construction claims. *Cost Engineering Journal, 46*(6), 16-20.

Saxena, A., (2008). *Enterprise contract management*. City: J. Ross Publishing.

Schieg, M. (2007). Post-mortem analysis on the analysis and evaluation of risk in construction project management. *Journal of Business Economic and Management, VIII* (2), 145-153.

Schleifer, C. T., (1990). *Construction contractors' survival guide*. New York: Wiley.

Shenhar, J. A., Levy, O., & Dvir, D. (1997). Mapping the dimension of project success. *Project Management Journal*, *28*(2), 5-13.

Shofoluwe, M., & Bogale, T. (2010). An investigative study of risk management practices of major U.S. contractors. The 7th International Symposium on
Risk Management and Cyber-Informatics: RMCI 2010 Orlando, Florida, USA. Retrieved from
http://www.iiis.org/CDs2010/CD2010SCI/RMCI_2010/PapersPdf/RA501MH.pdf

Small Business Administration. (2009). *The small business economy: A report to the President*. United States Government Printing Office. Washington: 2009

Smith, N., Merna, T., & Jobling, P. (2006). *Managing risk*. City: Blackwell Publishing Professional.

Spitz, M. G., Niles, L. F., & Adler, J. T. (2006). *TCRP Synthesis 69: Web-based Survey Techniques*. Transportation Research Board. National Research Council, Washington D.C

State of Florida Department of Transportation (nd). Major construction projects on Florida's Highways. Retrieved from http://www.dot.state.fl. us/publicinformationoffice/moreDOT/majorprojects.shtm

State of California Department of Transportation - Caltrans. (nd). Statement of ongoing contract reports, Retrieved from http://www.dot.ca.gov/hq/construc/statement.html

State of California Department of Transportation - Caltrans. (2008). Field guide to partnering on

Caltrans construction projects. Retrieved from

http://www.dot.ca.gov/hq/construc/Partnering_Fieldguide.pdf

State of California Department of Transportation - Caltrans. (2007). Caltrans project

management handbook. Retrieved from

http://www.dot.ca.gov/hq/projmgmt/documents/pmhb_5thed.pdf

State of California Department of Transportation - Caltrans. (2006). Time related overhead and

mobilization. Retrieved from http://sv08data.dot.ca.gov/contractcost/memo.pdf

State of California Department of Transportation - Caltrans. (2006). Decision document: Major

construction contract advertisement timeframes. Retrieved from

http://www.dot.ca.gov/hq/construc/GoCalifornia/closeout_reports/B1-c-2attach.pdf

State of California Department of Transportation - Caltrans. (1997). Subcontracting-contract

compliance. Retrieved from http://www.dot.ca.gov/hq/LocalPrograms/procrev/pr96-

05.pdf

State of Illinois Department of Transportation. (2007). Construction inspectors checklist for

contract administration. Retrieved from

http://www.dot.il.gov/const/curpdf/c_contractadministration.pdf

State of Kentucky Department of Highways. (2010). Contractor's performance report. Retrieved

form http://transportation.ky.gov/Organizational-Resources/Forms/TC%2014-19.pdf

State of Louisiana DOT and Development. (2011). Construction contract administration.

Retrieved from

http://www.dotd.la.gov/construction/Contract_Administration_Manual_May_2011.pdf

State of Minnesota Department of Transportation (2009). Contract administration manual.

Retrieved from http://www.dot.state.mn.us/const/tools/docs/cam.pdf

State of Texas Department of Transportation (2007). Construction contract administration

manual. Retrieved from http://onlinemanuals.txdot.gov/txdotmanuals/cah/cah.pdf

State of Utah Department of Transportation. (2004) Contractor Performance Rating Form.

Retrieved from http://www.udot.utah.gov/main/uconowner.gf?n=200510270725131

Sturgill, E. R. & Vorster, C. M. (2006). Visually improving construction contract administration.

Transportation Research Board: Journal of The Transportation Research Board. Vol.

1946, p12-21

Tichacek, L. R. (2006). Effective cost management: Back to the basics. *Cost Engineering

Journal*, 48(3), 27-33.

The National Academies (2007). *Reducing construction cost: Uses of best dispute resolution

practices by project owners.* National Academic Press. Washington D.C

Toole, T. M. (2006). "Social Science Research Methods in Construction." Proceedings of the

2006 ASCE-CIB Leadership in Construction Conference, May 2006.

U.S. Government Accountability Office (GAO). (2006). *Contract management: DOD

vulnerabilities to contracting fraud, waste, and abuse.* GAO. Washington DC.

U.S. Bureau of Economic Analysis. (2011). Gross domestic product by industry account.

Retrieved from http://www.bea.gov/industry/

U.S. Army Corps of Engineers (2002). *Contractor's Guide to Contract Administration.*
Retrieved from http://www.rmssupport.com/datafiles%5CContractors%20Guide.PDF

U.S. Census Bureau. (2011). Value of construction put in place. Retrieved from
http://www.census.gov/const/www/totpage.html

U.S. Merit Systems Protection Board. (2005). Contracting officer representatives: Managing the
government's technical experts to Achieve Positive Contract Outcome. U.S Merit
Systems Protection Board. Washington DC.

U.S. DOE. (2010). *Staffing Guide for Project Management.* DOE G 413.3-19

U.S. Department of Energy (2006). *The Guide for Contract Management Planning.* DOE,
Washington, D.C

U.S. Department of Energy. (2008). *Special Report on Managing Challenges at the Department
of Energy.* DOE /IG 0808

U.S. Department of Energy. (2008). *Root Cause Analysis Contract and Project Management:
Corrective Action Plan.* DOE, Washington, D.C

Vanden Bosche, R. P., (1981). The Cost Engineer's role in construction claims preparation and
prevention. *AACE Transactions*, Morgantown, W.V. D7.1-11

Visser, M. P., & Chermark, T. J. (2009). Perceptions of the relationship between scenario
planning and firm performance: A qualitative study. *Future*, 41(2009), p581-592

Walker, A. (2007). *Project management in construction.* Fifth Edition. City: Wiley-Blackwell.

Walker, D. H. T. (1995). An investigation into construction time performance. *Construction
Management and Economics, 13*(3), 263-274

Wegelius-Lehtonen, T. (2001). Performance measurement in construction logistics. *International Journal of Production Economics*, *69*(2001) 69, p107-116

Wyk, R. J. V. (2002). Technology: A fundamental structure? *Knowledge, Technology, & Policy*, *15*(3), 14-35.

Yau, N. (2011). *Visualize this: the FlowingData guide to design, visualization, and statistics*. Indianapolis, Ind: Wiley Pub.

Yeung, J. F. Y., Chan, A.P.C., & Chan, D.W.M. (2007). Development of partnering performance index (PPI) for construction projects in Hong Kong: a Delphi study. *Construction Management and Economics* (December 2007) 25, 1219–1237

Zazaian, M., (2006). Contract administration: A strategic business process. *Contract Management*, *46*(5), 48-52.

APPENDIX A: QUESTIONNAIRE

General Information

G1. Enter Project Name and Project Number

G2. What was the Project Value at Contract Award?

G3. What is the Current Project Value (Project Value at Contract Award Plus All Executed Change Orders)?

G4. Was Contract Competitively Bid and Awarded to the Lowest Bidder (Bid-Build)?

YES   NO

G5.

Indicate the Average Cycle Time from Discovery of Change to Formal Execution of Change Order. Discovery of change refers to when owner sent request for change or when contractor sent notice of change. Execution refers to when change order was formally signed by owner and contractor

	Total Number of Change Orders Executed (signed) at Each Period in Time	Average Length of Time from Discovery of Change to Execution of Formal Change Order
	Total Executed	Average Duration (in months) from Discovery of Change to Execution
JANUARY TO SEPTEMBER 2011		
JANUARY TO DECEMBER 2010		

Q1: Where partnering is encouraged or required, and implemented on a project, are additional partnering meetings and workshops held to maintain partnering relationships?

YES   NO   Not Applicable

Q2: Are significant contributors from the Contractor and Owner present at all partnering meetings and workshops?

YES   NO   Not Applicable

Q3: Does Contractor furnish competent supervisors to direct performance of work in accordance with the contract provisions?

YES   NO   Not Applicable

Q4: Does Owner furnish competent representatives to provide directions and make decisions on contract provisions?

YES   NO   Not Applicable

Q5: Are "us vs. them" attitudes discouraged by providing a contracting environment of shared trust, equity, and commitment?

YES   NO   Not Applicable

Q6: Does Owner provide periodic evaluation of contractor's rate of progress of the work?

YES   NO   Not Applicable

○   ○   ○

Q7: Does the contract provide for applicable labor, material, equipment, and subcontract

markups (overhead and profit) for extra work paid by force account method?

YES   NO   Not Applicable

○   ○   ○

Q8: Does contract allow time-related overhead (TRO) costs to include field and home-office

overhead for an increase in the time required to complete the work?

YES   NO   Not Applicable

○   ○   ○

Q9: For all dispute or potential claims where resolution at the project level were unsuccessful,

were dispute resolution members convened quickly to review the issues?

YES   NO   Not Applicable

○   ○   ○

Q10: Is professionally facilitated partnering encouraged or required by contracts?

YES   NO   Not Applicable

○   ○   ○

Q11: In compliance with the requirement on false statement concerning Highway Projects, are

notice on 18. U. S.C 1020 posted on project regarding false claim?

YES   NO   Not Applicable

◦   ◦   ◦

Q12: When changed condition is materially different from that of the original contract, are

Contractor's submitted change order proposals based on pre-established remedies/provisions?

YES   NO   Not Applicable

◦   ◦   ◦

Q13: Regarding payment withholding, has Contractor always been in compliance and portion of

Contractor's progress payment has not been withheld?

YES   NO   Not Applicable

◦   ◦   ◦

Q14: Do Contractor's change order proposals exclude costs for claim consultant and legal fees?

YES   NO   Not Applicable

◦   ◦   ◦

Q15: Are requirements for final inspection and acceptance of work items, and acceptance of

contract well-defined in the contract provisions?

YES   NO   Not Applicable

◦   ◦   ◦

Q16: Does Contractor have fewer than two (2) unresolved notification from the Owner on non-

compliance to certified payroll requirement?

YES	NO	Not Applicable
◌	◌	◌

Q17: Are disadvantaged business enterprise (DBE) records and reports submitted monthly to show compliance and good faith efforts?

YES	NO	Not Applicable
◌	◌	◌

Q18: Are applicable contract administration provisions (legal relations, changes, dispute resolutions, payments, progress schedule) clearly outline in the standard specifications and/or special provisions?

YES	NO	Not Applicable
◌	◌	◌

Q19: Has Contractor always maintained lower contract administration team turnover rate?

YES	NO	Not Applicable
◌	◌	◌

Q20: Is Owner dedicated design and engineering team available to promptly review and address omissions, errors and additions to contract plans and specifications?

YES	NO	Not Applicable
◌	◌	◌

Q21: Regarding extent of changes to the contract, are total number of (new items plus original items that were materially changed) extra work fewer than 10% of original work items?

YES	NO	Not Applicable

Q22: Does Contractor aggregate costs of labor used in the direct performance of change order item of work?

YES   NO   Not Applicable

Q23: Regarding proper documentation, is Contractor effective at supporting changes by providing required documentation to prove entitlement?

YES   NO   Not Applicable

Q24: Does Contractor aggregate project costs based on original scope, changed quantities, change in character of work, extra work, overhead, subcontract work, and potential changes?

YES   NO   Not Applicable

Q25: For extra work paid by force account, arc Contractors daily force account work report sheets prepared and signed on a daily basis?

YES   NO   Not Applicable

Q26: Are Contractor's competent supervisors quickly assigned to manage change orders work as soon as they are encountered?

YES   NO   Not Applicable

Q27: Do Contractor's representatives that attend dispute resolution meetings have knowledge of and authority to make decisions on the issues addressed?

YES   NO   Not Applicable

Q28: Is the Owner prompt at payment of undisputed progress payment request?

YES   NO   Not Applicable

Q29: Did Contractor assign competent contract administration team to the project at the start of project instead of waiting until changes are encountered?

YES   NO   Not Applicable

Q30: Regarding "Buy America" requirement has Contractor always submitted a certificate of compliance for steel and iron materials or appropriate waiver documentation?

YES   NO   Not Applicable

Q31: Has Contractor been in compliance of federal, state, or local laws?

YES   NO   Not Applicable

Q32: When applicable, is Contractor's claim on lost labor productivity supported by similar item of work (measured mile) that was not impacted?

YES   NO   Not Applicable

Q33: Are update schedule submitted monthly, current and no more than one (1) month behind?

YES   NO   Not Applicable

Q34: Do all time impact analyses (TIA) submitted by Contractor meet specified TIA preparation requirements?

YES   NO   Not Applicable

Q35: Did Contractor submit the baseline schedule by the due date?

YES   NO   Not Applicable

Q36: Is Contractor prompt at sending notice of potential claim (NOPC) to owner?

YES   NO   Not Applicable

Q37: Has Contractor always notified Owner on 1) physical conditions differing materially from contract document or job site examination and 2) physical conditions of unusual nature?

YES   NO   Not Applicable

Q38: Does Contractor prepared project schedule contain work breakdown structure (WBS) or identification codes for filtering, aggregating, and organizing activities?

YES   NO   Not Applicable

Q39: For a contract with a total bid of $25 million or greater, are Contractor's contract administrators trained in practices related to partnering?

YES   NO   Not Applicable

Q40: Does Contractor use visual charting, graphing or other visual representation when communicating and supporting potential claims?

YES   NO   Not Applicable

APPENDIX B: EMAIL REQUEST SENT TO STATE DOT CONTACT PERSON, AND REQUESTING THEIR HELP IN COORDINATING THE DISTRIBUTION OF THE SURVEY

My name is George Okere, I am a doctoral candidate at Indiana State University. I am currently working on a dissertation project aimed at 1) ***finding explanatory variables of contract administration practices that significantly relate to contract administration performance of General Contractors on State and Federal DOT projects, and 2) constructing and testing a model to predict the contract administration performance of Contractors on State and Federal DOT projects based on their contract administration practices.*** This is a very important research work that will add new knowledge to contract administration practice, and I am currently working with various state DOTs to help distribute the research survey. I understand that your office can help coordinate the distribution of the web-based research survey to some XXDOT Resident Engineers working on current projects. I am looking for five to ten XXDOT projects to complete the online survey. The selection criteria for projects to include in the research are as follows-

1. Project is a bridge and/or road project

2. Project is one year or more into construction

3. Project has encountered and executed some change orders

4. Project delivery method is design-bid-build (competitively bid project)

Thanks for your time, and I look forward to your response

Thanks

George Okere

APPENDIX C: EMAIL INVITATION LETTER SENT TO STATE DOT CONTACT PERSON FOR DISTRIBUTION TO RESIDENT ENGINEERS OF PROJECTS THAT MEET SET CRITERIA

Dear Participant,

An Investigative Study of Contract Administration Practices of General Contractors on Federal and State DOT Projects

You are being invited to participate in a research study about contract administration practices of contractors on federal and state DOT projects as it relates to contractors' ability to consistently achieve contract administration objectives. This study is being conducted by George O. Okere with Dr. Lee Ellingson as the faculty sponsor (dissertation committee chairman), from the College of Technology at Indiana State University. This study is being conducted as part of a dissertation project.

You were selected as a possible participant in this study because you have been identified as a project management practitioner involved in managing contracts for state DOTs, and would be interested in advancing the body of knowledge in contract administration, hence the request to participation in this survey.

There are no known risks if you decide to participate in this research study. There are no costs to you for participating in the study. The information you provide will be used to identify, define, predict and control factors that are related to Contractors' inability to consistently achieve contract administration objectives. The questionnaire will take about 30 minutes to complete. The information collected may not benefit you directly, but the information learned in this study should provide more general benefits.

The study promises anonymity of participants because the anonymous link feature of Qualtrics was used to distribute this survey.

The survey is being sent to you by your agency's coordinator and contact person for distribution of this study and email address or name of participants are not required, and as such no personal information has been collected of participant or maintained by this study. Individuals from Institutional Review Board may inspect these records. Should the data be published, no individual information will be disclosed.

Your participation in this study is voluntary. <u>If you consent to participating in this study, please follow the link below to the survey.</u>
<u>*If you DO NOT consent to participating in this study, please close this window to exit the study.*</u> *You are free to decline to answer any particular question you do not wish to answer for any reason.*

This study does not claim superiority, safety, or effectiveness of procedures, interventions, devices, or any other materials used in this study

Also note that there will be no future email contacts or an opt-out message that permits individuals to have their names removed from any future mailings.

Feel free to delete the email message that originated the contact

If you have any questions about the study, please contact *George O. Okere, 8533 Giverny Cir, Antelope, CA 95843, Phone number: 916 721-1680, and email address:* gokere@sycamores.indstate.edu

If you have any questions about your rights as a research subject or if you feel you've been placed at risk, you may contact the Indiana State University Institutional Review Board (IRB) by mail at Indiana State University, Office of Sponsored Programs, Terre Haute, IN, 47809, by phone at (812) 237-8217, or by e-mail at irb@indstate.edu.

Please note that the deadline to complete this survey is April 30th, 2012

Follow this link to the Survey:
${l://SurveyLink?d=Take the Survey}

Or copy and paste the URL below into your internet browser:
${l://SurveyURL}

Follow the link to opt out of future emails:
${l://OptOutLink?d=Click here to unsubscribe}

APPENDIX D: IRB APPROVAL LETTER

Indiana State
University
More. From day one.

Institutional Review Board

Terre Haute, Indiana 47809
812-237-3092
Fax: 812-237-3092

DATE: October 14, 2011

TO: George Okere
FROM: Indiana State University Institutional Review Board

STUDY TITLE: [259330-1] An Investigative Study of Contract Administration Practices of
                     General Contractors on Federal and State DOT Projects
IRB REFERENCE #:   12-024
SUBMISSION TYPE: New Project

ACTION: DETERMINATION OF EXEMPT STATUS
DECISION DATE: October 14, 2011

REVIEW CATEGORY: Exemption category #2, 5

Thank you for your submission of New Project materials for this research study. The Indiana State
University Institutional Review Board has determined this project is EXEMPT FROM IRB REVIEW
according to federal regulations (45 CFR 46). You do not need to submit continuation requests or a
completion report. Should you need to make modifications to your protocol or informed consent forms that
do not fall within the exempt categories, you will have to reapply to the IRB for review of your modified
study.

Internet Research: You are using an internet platform to collect data on human subjects. Although your
study is exempt from IRB review, ISU has specific policies about internet research that you should follow
to the best of your ability and capability. Please review Section L. on Internet Research in the IRB Policy
Manual.

Informed Consent: All ISU faculty, staff, and students conducting human subjects research within the
"exempt" category are still ethically bound to follow the basic ethical principles of the Belmont Report:
a) respect for persons; 2) beneficence; and 3) justice. These three principles are best reflected in the
practice of obtaining informed consent.

If you have any questions, please contact Thomas Steiger within IRBNet by clicking on the study title on
the "My Projects" screen and the "Send Project Mail" button on the left side of the "New Project Message"
screen. I wish you well in completing your study.

- 1 -

APPENDIX E: SAMPLE OF CONTRACTOR PERFORMANCE ASSESSMENT FORM (DD FORM 2626)

FOR OFFICIAL USE ONLY *(WHEN COMPLETED)*

PERFORMANCE EVALUATION (CONSTRUCTION)

1. CONTRACT NUMBER

2. CEC NUMBER

IMPORTANT: Be sure to complete Part III - Evaluation of Performance Elements on reverse.

PART I - GENERAL CONTRACT DATA

3. TYPE OF EVALUATION *(X one)* — INTERIM *(List percentage ___%)* — FINAL — AMENDED

4. TERMINATED FOR DEFAULT

5. CONTRACTOR *(Name, Address, and ZIP Code)*

6.a. PROCUREMENT METHOD *(X one)* — SEALED BID — NEGOTIATED

b. TYPE OF CONTRACT *(X one)* — FIRM FIXED PRICE — COST REIMBURSEMENT — OTHER *(Specify)*

7. DESCRIPTION AND LOCATION OF WORK

8. TYPE AND PERCENT OF SUBCONTRACTING

9. FISCAL DATA ▶ | a. AMOUNT OF BASIC CONTRACT $ | b. TOTAL AMOUNT OF MODIFICATIONS $ | c. LIQUIDATED DAMAGES ASSESSED $ | d. NET AMOUNT PAID CONTRACTOR $

10. SIGNIFICANT DATES ▶ | a. DATE OF AWARD | b. ORIGINAL CONTRACT COMPLETION DATE | c. REVISED CONTRACT COMPLETION DATE | d. DATE WORK ACCEPTED

PART II - PERFORMANCE EVALUATION OF CONTRACTOR

11. OVERALL RATING *(X appropriate block)* — OUTSTANDING — ABOVE AVERAGE — SATISFACTORY — MARGINAL — UNSATISFACTORY *(Explain in Item 20 on reverse)*

12. EVALUATED BY
a. ORGANIZATION *(Name and Address (Include ZIP Code))*
b. TELEPHONE NUMBER *(Include Area Code)*
c. NAME AND TITLE
d. SIGNATURE
e. DATE

13. EVALUATION REVIEWED BY
a. ORGANIZATION *(Name and Address (Include ZIP Code))*
b. TELEPHONE NUMBER *(Include Area Code)*
c. NAME AND TITLE
d. SIGNATURE
e. DATE

14. AGENCY USE *(Distribution, etc.)*

DD FORM 2626, JUN 94

EXCEPTION TO SF 1420 APPROVED BY GSA/IRMS 6-94

163

FOR OFFICIAL USE ONLY *(WHEN COMPLETED)*

PART III - EVALUATION OF PERFORMANCE ELEMENTS

N/A = NOT APPLICABLE   O = OUTSTANDING   A = ABOVE AVERAGE   S = SATISFACTORY   M = MARGINAL   U = UNSATISFACTORY

15. QUALITY CONTROL	N/A	O	A	S	M	U	16. EFFECTIVENESS OF MANAGEMENT	N/A	O	A	S	M	U
a. QUALITY OF WORKMANSHIP							a. COOPERATION AND RESPONSIVENESS						
b. ADEQUACY OF THE CQC PLAN							b. MANAGEMENT OF RESOURCES/ PERSONNEL						
c. IMPLEMENTATION OF THE CQC PLAN							c. COORDINATION AND CONTROL OF SUBCONTRACTOR(S)						
d. QUALITY OF QC DOCUMENTATION							d. ADEQUACY OF SITE CLEAN-UP						
e. STORAGE OF MATERIALS							e. EFFECTIVENESS OF JOB-SITE SUPERVISION						
f. ADEQUACY OF MATERIALS													
g. ADEQUACY OF SUBMITTALS							f. COMPLIANCE WITH LAWS AND REGULATIONS						
h. ADEQUACY OF QC TESTING							g. PROFESSIONAL CONDUCT						
i. ADEQUACY OF AS-BUILTS							h. REVIEW/RESOLUTION OF SUBCONTRACTOR'S ISSUES						
j. USE OF SPECIFIED MATERIALS													
k. IDENTIFICATION/CORRECTION OF DEFICIENT WORK IN A TIMELY MANNER							i. IMPLEMENTATION OF SUBCONTRACTING PLAN						
17. TIMELY PERFORMANCE							**18. COMPLIANCE WITH LABOR STANDARDS**						
a. ADEQUACY OF INITIAL PROGRESS SCHEDULE							a. CORRECTION OF NOTED DEFICIENCIES						
b. ADHERENCE TO APPROVED SCHEDULE							b. PAYROLLS PROPERLY COMPLETED AND SUBMITTED						
c. RESOLUTION OF DELAYS							c. COMPLIANCE WITH LABOR LAWS AND REGULATIONS WITH SPECIFIC ATTENTION TO THE DAVIS-BACON ACT AND EEO REQUIREMENTS						
d. SUBMISSION OF REQUIRED DOCUMENTATION													
e. COMPLETION OF PUNCHLIST ITEMS							**19. COMPLIANCE WITH SAFETY STANDARDS**						
f. SUBMISSION OF UPDATED AND REVISED PROGRESS SCHEDULES							a. ADEQUACY OF SAFETY PLAN						
							b. IMPLEMENTATION OF SAFETY PLAN						
g. WARRANTY RESPONSE							c. CORRECTION OF NOTED DEFICIENCIES						

20. REMARKS *(Explanation of unsatisfactory evaluation is required. Other comments are optional. Provide facts concerning specific events or actions to justify the evaluation. These data must be in sufficient detail to assist contracting officers in determining the contractor's responsibility. Continue on separate sheet(s), if needed.)*

DD FORM 2626 (BACK), JUN 94

APPENDIX F: CONTRACT ADMINISTRATION PRACTICES AND THE DEGREE WITH WHICH THEY ARE IN PRACTICE BY THE SAMPLED STATE DOT PROJECTS IN THE U.S

Contract Administration Practices that are Expected to Improve Contract Administration Performance of Contractors on federal and state DOT Projects in the U.S.	Response Options Provided to Respondents	Number and Percentage of Respondents Per Response Options	Number of Respondents that Indicated a Response Option	Number of Respondent Without a Response Option	Percentage of Respondents Without a Response Option
(Q1)Where partnering is encouraged or required, and implemented on a project, are additional partnering meetings and workshops held to maintain partnering relationships?	YES	36 42%			
	NO	25 29%			
	Not Applicable	25 29%	86	0	0%

Contract Administration Practices that are Expected to Improve Contract Administration Performance of Contractors on federal and state DOT Projects in the U.S.	Response Options Provided to Respondents	Number and Percentage of Respondents Per Response Options	Number of Respondents that Indicated a Response Option	Number of Respondent Without a Response Option	Percentage of Respondents Without a Response Option
(Q2)Are significant contributors from the Contractor and Owner present at all partnering meetings and workshops?	YES	52 60%			
	NO	3 3%			
	Not Applicable	31 36%	86	0	0%
(Q3)Does Contractor furnish competent supervisors to direct performance of work in accordance with the contract provisions?	YES	78 91%			
	NO	7 8%			
	Not Applicable	0 0%	85	1	1%

Contract Administration Practices that are Expected to Improve Contract Administration Performance of Contractors on federal and state DOT Projects in the U.S.	Response Options Provided to Respondents	Number and Percentage of Respondents Per Response Options	Number of Respondents that Indicated a Response Option	Number of Respondent Without a Response Option	Percentage of Respondents Without a Response Option
	YES	82			
(Q4)Does Owner furnish competent representatives to provide directions and make decisions on contract provisions?		95%			
	NO	3 3%			
	Not Applicable	1 1%	86	0	0%
	YES	79 92%			
(Q5)Are us vs. them attitudes discouraged by providing a contracting environment of shared trust, equity, and commitment?					
	NO	2 2%			
	Not Applicable	4 5%	85	1	1%

Contract Administration Practices that are Expected to Improve Contract Administration Performance of Contractors on federal and state DOT Projects in the U.S.	Response Options Provided to Respondents	Number and Percentage of Respondents Per Response Options	Number of Respondents that Indicated a Response Option	Number of Respondent Without a Response Option	Percentage of Respondents Without a Response Option
(Q6)Does Owner provide periodic evaluation of contractor's rate of progress of the work?	YES	70 81%			
	NO	15 17%			
	Not Applicable	1 1%	86	0	0%
(Q7)Does the contract provide for applicable labor, material, equipment, and subcontract markups (overhead and profit) for extra work paid by force account method?	YES	81 94%			
	NO	1 1%			
	Not Applicable	4 5%	86	0	0%

Contract Administration Practices that are Expected to Improve Contract Administration Performance of Contractors on federal and state DOT Projects in the U.S.	Response Options Provided to Respondents	Number and Percentage of Respondents Per Response Options	Number of Respondents that Indicated a Response Option	Number of Respondent Without a Response Option	Percentage of Respondents Without a Response Option
(Q8)Is time related overhead (TRO) provision in the contract?	YES	28 33%			
	NO	45 52%			
	Not Applicable	7 8%	80	6	7%
(Q9)For all dispute or potential claims where resolution at the project level were unsuccessful, were dispute resolution members convened quickly to review the issues?	YES	31 36%			
	NO	8 9%			
	Not Applicable	45 52%	84	2	2%
(Q10)Is professionally facilitated partnering encouraged or required by contracts?	YES	36 42%			
	NO	38 44%			
	Not	12			

Contract Administration Practices that are Expected to Improve Contract Administration Performance of Contractors on federal and state DOT Projects in the U.S.	Response Options Provided to Respondents	Number and Percentage of Respondents Per Response Options	Number of Respondents that Indicated a Response Option	Number of Respondent Without a Response Option	Percentage of Respondents Without a Response Option
	Applicable	14%	86	0	0%
(Q11)In compliance with the requirement on false statement concerning Highway Projects, are notice on 18. U. S.C 1020 posted on project regarding false claim?	YES	47 55%			
	NO	16 19%			
	Not Applicable	10 12%	73	13	15%
(Q12)When changed condition is materially different from that of the original contract, are Contractor's submitted change order prices based on pre-established remedies/provisions?	YES	43 50%			
	NO	32 37%			
	Not Applicable	10 12%	85	1	1%

Contract Administration Practices that are Expected to Improve Contract Administration Performance of Contractors on federal and state DOT Projects in the U.S.	Response Options Provided to Respondents	Number and Percentage of Respondents Per Response Options	Number of Respondents that Indicated a Response Option	Number of Respondent Without a Response Option	Percentage of Respondents Without a Response Option
(Q13)Regarding payment withholding, has Contractor always been in compliance and portion of Contractor's progress payment has not been withheld?	YES	45 52%			
	NO	25 29%			
	Not Applicable	14			
		16%	84	2	2%
(Q14)Do Contractor's change order proposals exclude costs for claim consultant and legal fees?	YES	60 70%			
	NO	7 8%			
	Not Applicable	17			
		20%	84	2	2%

Contract Administration Practices that are Expected to Improve Contract Administration Performance of Contractors on federal and state DOT Projects in the U.S.	Response Options Provided to Respondents	Number and Percentage of Respondents Per Response Options	Number of Respondents that Indicated a Response Option	Number of Respondent Without a Response Option	Percentage of Respondents Without a Response Option
(Q15)Are requirements for final inspection and acceptance of work items, and acceptance of contract well-defined in the contract provisions?	YES	84 98%			
	NO	2 2%			
	Not Applicable	0 0%	86	0	0%
(Q16)Does Contractor have fewer than two (2) unresolved notification from the Owner on non-compliance to certified payroll requirement?	YES	59 69%			
	NO	18 21%			
	Not Applicable	7 8%	84	2	2%

Contract Administration Practices that are Expected to Improve Contract Administration Performance of Contractors on federal and state DOT Projects in the U.S.	Response Options Provided to Respondents	Number and Percentage of Respondents Per Response Options	Number of Respondents that Indicated a Response Option	Number of Respondent Without a Response Option	Percentage of Respondents Without a Response Option
(Q17)Are disadvantaged business enterprise (DBE) records and reports submitted monthly to show compliance and good faith efforts?	YES	60 70%			
	NO	13 15%			
	Not Applicable	10 12%	83	3	3%
(Q18)Are applicable contract administration provisions (legal relations, changes, dispute resolutions, payments, progress schedule) clearly outline in the standard specifications and/or special provisions?	YES	85 99%			
	NO	1 1%			
	Not Applicable	0 0%	86	0	0%

Contract Administration Practices that are Expected to Improve Contract Administration Performance of Contractors on federal and state DOT Projects in the U.S.	Response Options Provided to Respondents	Number and Percentage of Respondents Per Response Options	Number of Respondents that Indicated a Response Option	Number of Respondent Without a Response Option	Percentage of Respondents Without a Response Option
(Q19)Has Contractor always maintained lower contract administration team turnover rate?	YES	47 55%			
	NO	11 13%			
	Not Applicable	20 23%	78	8	9%
(Q20)Is Owner dedicated design and engineering team available to promptly review and address omissions, errors and additions to contract plans and specifications?	YES	74 86%			
	NO	9 10%			
	Not Applicable	3 3%	86	0	0%

Contract Administration Practices that are Expected to Improve Contract Administration Performance of Contractors on federal and state DOT Projects in the U.S.	Response Options Provided to Respondents	Number and Percentage of Respondents Per Response Options	Number of Respondents that Indicated a Response Option	Number of Respondent Without a Response Option	Percentage of Respondents Without a Response Option
(Q21)Regarding extent of changes to the contract, is extra work fewer than 10 of original contract?	YES	66 77%			
	NO	15 17%			
	Not Applicable	4 5%	85	1	1%
(Q22)Does Contractor aggregate costs of labor used in the direct performance of change order item of work?	YES	48 56%			
	NO	15 17%			
	Not Applicable	16 19%	79	7	8%

Contract Administration Practices that are Expected to Improve Contract Administration Performance of Contractors on federal and state DOT Projects in the U.S.	Response Options Provided to Respondents	Number and Percentage of Respondents Per Response Options	Number of Respondents that Indicated a Response Option	Number of Respondent Without a Response Option	Percentage of Respondents Without a Response Option
(Q23)Regarding proper documentation, is Contractor effective at supporting changes by providing required documentation to prove entitlement?	YES	69 80%			
	NO	15 17%			
	Not Applicable	2 2%	86	0	0%
(Q24)Does Contractor aggregate project costs based on original scope, changed quantities, change in character of work, extra work, overhead, subcontract work, and potential changes?	YES	57 66%			
	NO	12 14%			
	Not Applicable	8 9%	77	9	10%

Contract Administration Practices that are Expected to Improve Contract Administration Performance of Contractors on federal and state DOT Projects in the U.S.	Response Options Provided to Respondents	Number and Percentage of Respondents Per Response Options	Number of Respondents that Indicated a Response Option	Number of Respondent Without a Response Option	Percentage of Respondents Without a Response Option
(Q25)For extra work paid by force account, are Contractors daily force account work report sheets prepared and signed on a daily basis?	YES	54 63%			
	NO	3 3%			
	Not Applicable	29 34%	86	0	0%
(Q26)Are Contractor's competent supervisors quickly assigned to manage change orders work as soon as they are encountered?	YES	68 79%			
	NO	11 13%			
	Not Applicable	7 8%	86	0	0%

Contract Administration Practices that are Expected to Improve Contract Administration Performance of Contractors on federal and state DOT Projects in the U.S.	Response Options Provided to Respondents	Number and Percentage of Respondents Per Response Options	Number of Respondents that Indicated a Response Option	Number of Respondent Without a Response Option	Percentage of Respondents Without a Response Option
(Q27)Do Contractor's representatives that attend dispute resolution meetings have knowledge of and authority to make decisions on the issues addressed?	YES	63 73%			
	NO	4 5%			
	Not Applicable	19 22%	86	0	0%
(Q28)Is the Owner prompt at payment of undisputed progress payment request?	YES	83 97%			
	NO	2 2%			
	Not Applicable	1 1%	86	0	0%

Contract Administration Practices that are Expected to Improve Contract Administration Performance of Contractors on federal and state DOT Projects in the U.S.	Response Options Provided to Respondents	Number and Percentage of Respondents Per Response Options	Number of Respondents that Indicated a Response Option	Number of Respondent Without a Response Option	Percentage of Respondents Without a Response Option
(Q29)Did Contractor assign competent contract administration team to the project at the start of project instead of waiting until changes are encountered?	YES	75 87%			
	NO	8 9%			
	Not Applicable	1 1%	84	2	2%
(Q30)Regarding "Buy America" requirement, has Contractor always submitted a certificate of compliance for steel and iron materials or appropriate waiver documentation?	YES	68 79%			
	NO	6 7%			
	Not Applicable	9 10%	83	3	3%

Contract Administration Practices that are Expected to Improve Contract Administration Performance of Contractors on federal and state DOT Projects in the U.S.	Response Options Provided to Respondents	Number and Percentage of Respondents Per Response Options	Number of Respondents that Indicated a Response Option	Number of Respondent Without a Response Option	Percentage of Respondents Without a Response Option
(Q31)Has Contractor been in compliance of federal, state, or local laws?	YES	82 95%			
	NO	1 1%			
	Not Applicable	1 1%	84	2	2%
(Q32)When applicable, is Contractor's claim on lost labor productivity supported by similar item of work (measured mile) that was not impacted?	YES	17 20%			
	NO	10 12%			
	Not Applicable	54 63%	81	5	6%
(Q33)Are update schedule submitted monthly, current and no more than one (1) month behind?	YES	49 57%			
	NO	28 33%			
	Not	8			

Contract Administration Practices that are Expected to Improve Contract Administration Performance of Contractors on federal and state DOT Projects in the U.S.	Response Options Provided to Respondents	Number and Percentage of Respondents Per Response Options	Number of Respondents that Indicated a Response Option	Number of Respondent Without a Response Option	Percentage of Respondents Without a Response Option
	Applicable	9%	85	1	1%
	YES	24 28%			
(Q34)Do all time impact analyses (TIA) submitted by Contractor meet specified TIA preparation requirements?					
	NO	11 13%			
	Not Applicable	44 51%	79	7	8%
	YES	70 81%			
(Q35)Did Contractor submit the baseline schedule by the due date?					
	NO	10 12%			
	Not Applicable	3 3%	83	3	3%
	YES	49 57%			
(Q36)Is Contractor prompt at sending notice of potential claim (NOPC) to owner?					
	NO	9 10%			
	Not Applicable	26			

Contract Administration Practices that are Expected to Improve Contract Administration Performance of Contractors on federal and state DOT Projects in the U.S.	Response Options Provided to Respondents	Number and Percentage of Respondents Per Response Options	Number of Respondents that Indicated a Response Option	Number of Respondent Without a Response Option	Percentage of Respondents Without a Response Option
		30%	84	2	2%
(Q37)Has Contractor always notified Owner on 1) physical conditions differing materially from contract document or job site examination and 2) physical conditions of unusual nature?	YES	62 72%			
	NO	12 14%			
	Not Applicable	10			
		12%	84	2	2%
(Q38)Does Contractor prepared project schedule contain work breakdown structure (WBS) or identification codes for filtering, aggregating, and organizing activities?	YES	51 59%			
	NO	20 23%			
	Not Applicable	13			
		15%	84	2	2%

Contract Administration Practices that are Expected to Improve Contract Administration Performance of Contractors on federal and state DOT Projects in the U.S.	Response Options Provided to Respondents	Number and Percentage of Respondents Per Response Options	Number of Respondents that Indicated a Response Option	Number of Respondent Without a Response Option	Percentage of Respondents Without a Response Option
(Q39)For a contract with a total bid of $10 million or greater, are Contractor's contract administrators trained in practices related to partnering?	YES	36 42%			
	NO	10 12%			
	Not Applicable	33 38%	79	7	8%
(Q40)Does Contractor use visual charting, graphing or other visual representation when communicating and supporting potential claims?	YES	17 20%			
	NO	34 40%			
	Not Applicable	31 36%	82	4	5%

APPENDIX G: AVERAGE DURATION TO RESOLVE A CHANGE, SCOPE VARIANCE, NUMBER AND RATE OF PRACTICES IN PLACE

Case ID	% of Change from Original Project Value	Total Number of Change Orders Executed in 2011	Total Number of Change Orders Executed in 2010	Average Length of Time (In Month) from Discovery of Change to Execution in 2011	Average Length of Time (In Month) from Discovery of Change to Execution in 2010	Combined Average Length of Time -2010 and 2011
R_bjaOdF9ZKJgq7SA	5.2%					
R_cZIQbzy8RYS9AqM	-25.3%					
		21.00	14.00	5.00	5.00	5.00
R_9ztGDmi4DeqdGNC	0.0%					
		9.00	6.00	1.00	1.00	1.00
R_eIC4qtsgdMXSQfO	2.0%					
		7.00	7.00	1.00	1.00	1.00
R_bEKzhm06R0xba8Q	-2.2%					
		10.00	7.00	1.00	1.00	1.00
R_77iD8xMuyCNpGSw	6.8%					
		12.00	23.00	1.00	1.00	1.00
R_5pOQSdIEcC9Go8k	1.0%					
		12.00	3.00			
R_bwj0jzwz67oo9IU	3.3%					
		2.00	12.00	1.00	3.00	2.00
R_74BuMltcGwLEs9m	10.0%					
		23.00		1.00		1.00
R_8j1neSJZOai4BU0	12.6%					
		47.00	6.00	1.00	1.00	1.00
R_d0hXo4F05FizONm	10.7%					
R_09i0VCFFc01VYLG	0.0%					
		11.00	3.00	1.00	1.00	1.00

Case ID	% of Change from Original Project Value	Total Number of Change Orders Executed in 2011	Total Number of Change Orders Executed in 2010	Average Length of Time (In Month) from Discovery of Change to Execution in 2011	Average Length of Time (In Month) from Discovery of Change to Execution in 2010	Combined Average Length of Time -2010 and 2011
R_9nsynjf1gxAKvxW	0.0%	12.00		2.00		2.00
R_a9vxAcaUIAp0W8c	5.1%	39.00	38.00	2.00	2.00	2.00
R_0lCOCM6Wmvp0gio	4.1%	30.00	30.00	2.00	2.00	2.00
R_bO40S00cImaUeiM	2.7%	70.00	60.00	1.00	2.00	1.50
R_bfNTaAOzicEmmdm	2.7%	11.00	9.00	1.00	1.00	1.00
R_0dZ6K0rDoUh0990	1.5%					
R_9z7EG0014ZO0eIk	7.0%	3.00				
R_0mVbCACxcnci0S0	10.3%	9.00	5.00	2.00	1.00	1.50
R_eJXRyQ0hsUNNAhK	33.9%		35.00			
R_5iiFkP05s9Lq89m	0.0%			3.00		3.00
R_00s0J9eWQPHlkJ6	4.7%	65.00	50.00	1.00	1.00	1.00
R_00qU0JMHbjCo910						
R_57JoCcWw7eqZI68	-8.4%	1.00		1.00		1.00
R_6J1YEgoo47Owwh6	1.8%	2.00		1.00		1.00
R_098rSvTVBu0zNWI	3.7%	4.00		1.00		1.00
R_1TtTE7u7XTcjqFm	10.5%					
R_1CflX0whac6Oeig	1.7%	4.00		3.00		3.00
R_5ax87b7M88v06wI	1.2%	2.00	2.00	2.00	1.00	1.50

Case ID	% of Change from Original Project Value	Total Number of Change Orders Executed in 2011	Total Number of Change Orders Executed in 2010	Average Length of Time (In Month) from Discovery of Change to Execution in 2011	Average Length of Time (In Month) from Discovery of Change to Execution in 2010	Combined Average Length of Time -2010 and 2011
R_efYOjGyGYllkenq	5.2%	16.00	19.00	3.00	3.00	3.00
R_ePXrtbu84QOy8u0	5.2%	16.00	19.00	3.00	3.00	3.00
R_cILgFHlM1gHqzdy	0.5%	6.00	6.00	1.00	1.00	1.00
R_cAzmkCXiRVTxzww	1.5%	13.00	13.00	1.00	1.00	1.00
R_9NAHtmvN4dUzxWs	14.4%	14.00	9.00	3.00	3.00	3.00
R_b1Qr0rZZ6epbhly	0.9%	21.00	7.00		1.00	1.00
R_ddnObNZSG7TvE9e	6.1%	3.00		2.00		2.00
R_a4zv6qFVmejHlPK	3.6%	12.00		5.00		5.00
R_000eLoxElVTFPso	0.0%					
R_0wnMpEv87HWDegk	5.5%	7.00		2.00		2.00
R_9XHCPKlL0narVkg	3.7%	19.00	4.00	1.00	1.00	1.00
R_55bsovRZTOFZgRS	16.8%	4.00	2.00	1.00	1.00	1.00
R_5upApIoYf6ZFnkU	10.9%	1.00	2.00	3.00	2.00	2.50
R_cTlFyPktekahcSo	7.0%	5.00		6.00		6.00
R_9pBzDL7K7UhJEVu	12.5%	9.00		1.00		1.00
R_0T9tkybSNAA79NG	-2.4%	12.00		1.00		1.00
R_0pXPDp6i1FXAzfm	22.9%	13.00	8.00			
R_ebdqt9OiuEpUp9i	0.0%					

Case ID	% of Change from Original Project Value	Total Number of Change Orders Executed in 2011	Total Number of Change Orders Executed in 2010	Average Length of Time (In Month) from Discovery of Change to Execution in 2011	Average Length of Time (In Month) from Discovery of Change to Execution in 2010	Combined Average Length of Time -2010 and 2011
R_0HF8w0E6XiZ8QS0	1.6%	4.00	3.00	6.00	1.00	3.50
R_afclGk4uw5LIozG	-2.7%	18.00		2.00		2.00
R_bdvqoj6pQShGvVG	-0.4%	15.00	4.00	1.00	1.00	1.00
R_0Pi0Foom61MyLsw	10.3%	7.00	11.00	3.00	3.00	3.00
R_cZU6NtqAqcGjmHG	4.8%	10.00	1.00	4.00	4.00	4.00
R_7TWHjtCzzj1MbGI	1.5%	14.00		1.00		1.00
R_beEr4v7YoazeGQQ	5.5%	3.00		6.00		6.00
R_6kWKfZ6pK8YMbDm	13.5%	16.00				
R_etv56VF6lGBgEza	-1.5%	5.00	11.00			
R_07C68fjypFDDUrO	6.6%	10.00		4.00		4.00
R_9AoeP0h010NW0Iw	56.4%	9.00	22.00	2.00	2.00	2.00
R_afSI9kIEo0DF7gM	3.8%	13.00		1.00		1.00
R_09vGjDj70XGUgQs	21.1%	2.00	1.00	3.00	3.00	3.00
R_eKWM5DfEXRS6Nog	3.1%	24.00	13.00	1.00	1.00	1.00
R_bJDW4SsVxsWA190	0.0%	1.00	1.00	2.00	3.00	2.50
R_9mZ5ZTebBQSFklS	0.0%					
R_6rgVpSHChf8Kp4E	2.1%	7.00		1.00		1.00
R_9zP7tGBNPK9fmkI	26.8%					

Case ID	% of Change from Original Project Value	Total Number of Change Orders Executed in 2011	Total Number of Change Orders Executed in 2010	Average Length of Time (In Month) from Discovery of Change to Execution in 2011	Average Length of Time (In Month) from Discovery of Change to Execution in 2010	Combined Average Length of Time -2010 and 2011
R_e5w17AXiQErpEmo	0.8%	9.00	2.00	1.00	1.00	1.00
R_0rfhpVZRoKyIP0I	10.1%	2.00		2.00		2.00
R_0lvkhmbo61ktzrC	-4.2%	26.00	10.00		1.00	1.00
R_eQJUcH6a9JSQfT6	0.0%	3.00		3.00		3.00
R_6PruhM0UxVltGjG	-2.0%	20.00	10.00	2.00	1.00	1.50
R_eFAG9ryOTReJlgE	-8.9%	3.00				
R_a96hrZLpZipU5LK	0.0%	14.00		2.00		2.00
R_1HXkwyW5C050ZA8	7.0%	54.00	24.00	2.00	2.00	2.00
R_0meUXfgewhAOUv0	1.2%	70.00	10.00	5.00	4.00	4.50
R_1ZjdR66ohqoxx88	6.2%	2.00		1.00		1.00
R_eXRG067eWdAcvfm	-2.2%	4.00	11.00	3.00	5.00	4.00
R_0w7y1UlDoSDceJ6	1.4%	4.00				
R_eR4e4NUvdnuGLjK	7.4%	1.00	1.00	1.00	1.00	1.00
R_0rTHXEDiC0HUqDa	10.1%	20.00	12.00	2.00	2.00	2.00
R_5Ac5v8gqw0keaP0	4.3%	6.00	3.00	7.00	8.00	7.50
R_9zPb7yIfN1LdJWc	5.9%	15.00		1.00		1.00
R_0hnD6yUdhcpvL1y		13.00		1.00		1.00

Case ID	% of Change from Original Project Value	Total Number of Change Orders Executed in 2011	Total Number of Change Orders Executed in 2010	Average Length of Time (In Month) from Discovery of Change to Execution in 2011	Average Length of Time (In Month) from Discovery of Change to Execution in 2010	Combined Average Length of Time -2010 and 2011
R_5dyPrp5rj94gAHq	1.2%					
		10.00	2.00	1.00	2.00	1.50
R_9HyawxnGJDH04EQ	0.0%					
		13.00		1.00		1.00
R_emw09JmPfRo9xre	0.7%					
		12.00	10.00	2.00	2.00	2.00
Percentage of Respondents Whose Project Experienced Change in Scope	90.5%					
Count	84	75.00	47.00	67.00	43.00	69.00
Mean	4.9%	13.81	11.94	2.12	2.02	2.07
Standard Deviation	9.5%	14.91	12.74	1.50	1.47	1.44

APPENDIX H: HYPOTHETICAL AGGREGATE AND CATEGORIES OF PRACTICES

Underlying Independent Variables and Corresponding Practices Measured					
Management attitude towards contract risks	Contract provisions for mitigating contract risks	Stability of scope definition and requirements	Contract administration infrastructure	Resource allocation strategy	Competency of contract administrators
(Q1)Where partnering is encouraged or required, and implemented on a project, are additional partnering meetings and workshops held to maintain partnering relationships?	(Q7)Does the contract provide for applicable labor, material, equipment, and subcontract markups (overhead and profit) for extra work paid by force account method?	(Q19)Has Contractor always maintained lower contract administration team turnover rate?	(Q22)Does Contractor aggregate costs of labor used in the direct performance of change order item of work?	(Q26)Are Contractor's competent supervisors quickly assigned to manage change orders work as soon as they are encountered?	(Q9)For all dispute or potential claims where resolution at the project level were unsuccessful, were dispute resolution members convened quickly to review the issues?
(Q2)Are significant contributors from the Contractor and Owner present at all partnering meetings and	(Q8)Is time related overhead (TRO) provision in the contract?	(Q20)Is Owner dedicated design and engineering team available to promptly review and address omissions, errors and additions to	(Q23)Regarding proper documentation, is Contractor effective at supporting changes by providing required documentation	(Q27)Do Contractor's representatives that attend dispute resolution meetings have knowledge	(Q11)In compliance with the requirement on false statement concerning Highway Projects, are notice on 18. U. S.C 1020 posted

Underlying Independent Variables and Corresponding Practices Measured					
Management attitude towards contract risks	Contract provisions for mitigating contract risks	Stability of scope definition and requirements	Contract administration infrastructure	Resource allocation strategy	Competency of contract administrators
workshops?		contract plans and specifications?	to prove entitlement?	of and authority to make decisions on the issues addressed?	on project regarding false claim?
(Q3)Does Contractor furnish competent supervisors to direct performance of work in accordance with the contract provisions?	(Q10)Is professionally facilitated partnering encouraged or required by contracts?	(Q21)Regarding extent of changes to the contract, is extra work fewer than 10 of original contract?	(Q24)Does Contractor aggregate project costs based on original scope, changed quantities, change in character of work, extra work, overhead, subcontract work, and potential changes?	(Q29)Did Contractor assign competent contract administration team to the project at the start of project instead of waiting until changes are encountered?	(Q12)When changed condition is materially different from that of the original contract, are Contractor's submitted change order prices based on pre-established remedies/provisions?
(Q4)Does Owner furnish competent representatives to provide directions and make decisions on contract provisions?	(Q15)Are requirements for final inspection and acceptance of work items, and acceptance of contract well-defined in the contract provisions?				(Q13)Regarding payment withholding, has Contractor always been in compliance and portion of Contractor's progress payment has not been withheld?

Underlying Independent Variables and Corresponding Practices Measured					
Management attitude towards contract risks	Contract provisions for mitigating contract risks	Stability of scope definition and requirements	Contract administration infrastructure	Resource allocation strategy	Competency of contract administrators
(Q5)Are us vs. them attitudes discouraged by providing a contracting environment of shared trust, equity, and commitment?	(Q18)Are applicable contract administration provisions (legal relations, changes, dispute resolutions, payments, progress schedule) clearly outline in the standard specifications and/or special provisions?				(Q14)Do Contractor's change order proposals exclude costs for claim consultant and legal fees?
(Q6)Does Owner provide periodic evaluation of contractor's rate of progress of the work?					(Q16)Does Contractor have fewer than two (2) unresolved notification from the Owner on non-compliance to certified payroll requirement?

Underlying Independent Variables and Corresponding Practices Measured					
Management attitude towards contract risks	Contract provisions for mitigating contract risks	Stability of scope definition and requirements	Contract administration infrastructure	Resource allocation strategy	Competency of contract administrators
(Q28)Is the Owner prompt at payment of undisputed progress payment request?					(Q17)Are disadvantaged business enterprise (DBE) records and reports submitted monthly to show compliance and good faith efforts? (Q25)For extra work paid by force account, are Contractors daily force account work report sheets prepared and signed on a daily basis?
					(Q30)Regarding "Buy America" requirement, has Contractor always submitted a certificate of compliance for steel and iron materials or appropriate waiver

Underlying Independent Variables and Corresponding Practices Measured					
Management attitude towards contract risks	Contract provisions for mitigating contract risks	Stability of scope definition and requirements	Contract administration infrastructure	Resource allocation strategy	Competency of contract administrators
					documentation?
					(Q31)Has Contractor been in compliance of federal, state, or local laws? (Q32)When applicable, is Contractor's claim on lost labor productivity supported by similar item of work (measured mile) that was not impacted? (Q33)Are update schedule submitted monthly, current and no more than one (1) month behind? (Q34)Do all time impact analyses (TIA) submitted by Contractor meet specified TIA preparation requirements?

Underlying Independent Variables and Corresponding Practices Measured					
Management attitude towards contract risks	Contract provisions for mitigating contract risks	Stability of scope definition and requirements	Contract administration infrastructure	Resource allocation strategy	Competency of contract administrators
					(Q35)Did Contractor submit the baseline schedule by the due date? (Q36)Is Contractor prompt at sending notice of potential claim (NOPC) to owner? (Q37)Has Contractor always notified Owner on 1) physical conditions differing materially from contract document or job site examination and 2) physical conditions of unusual nature?

Underlying Independent Variables and Corresponding Practices Measured					
Management attitude towards contract risks	Contract provisions for mitigating contract risks	Stability of scope definition and requirements	Contract administration infrastructure	Resource allocation strategy	Competency of contract administrators
					(Q38)Does Contractor prepared project schedule contain work breakdown structure (WBS) or identification codes for filtering, aggregating, and organizing activities?
					(Q39)For a contract with a total bid of $10 million or greater, are Contractor's contract administrators trained in practices related to partnering?
					(Q40)Does Contractor use visual charting, graphing or other visual representation when communicating and supporting potential claims?

APPENDIX I: TRANSFORMING DEPENDENT VARIABLE FOR STATISTICAL ANALYSIS

Average Cycle Time (In Months)- Collected Data	Performance Level –Recoded Data
1	7.5
1.5	7
2	6.5
2.5	6
3	5.5
3.5	5
4	4.5
4.5	4
5	3.5
5.5	3
6	2.5
6.5	2
7	1.5
7.5	1

APPENDIX J: DATA USED FOR INFERENTIAL

STATISTICS

Case #	Case ID	DV	Reverse Recoded DV	IV1	IV2	IV3	IV4	IV5	IV6
1	R_cZIQbzy8RYS9AqM	5	3.5	4	5	2	1	2	13
2	R_9ztGDmi4DeqdGNC	1	7.5	6	5	3	0	3	7
3	R_eIC4qtsgdMXSQfO	1	7.5	7	3	3	2	2	9
4	R_bEKzhm36R3xba8Q	1	7.5	5	4	3	3	3	11
5	R_77iD8xMuyCNpGSw	1	7.5	7	5	3	3	3	14
6	R_bwj0jzwz67oo9IU	2	6.5	7	4	3	3	3	12
7	R_74BuMltcGwLEs9m	1	7.5	7	4	3	2	3	12
8	R_8j1neSJZOai4BU0	1	7.5	4	5	1	1	3	10
9	R_39i2VCFFc31VYLG	1	7.5	6	4	3	2	3	13
10	R_9nsynjf1gxAKvxW	2	6.5	6	5	1	3	3	9
11	R_a9vxAcaUIAp2W8c	2	6.5	7	5	3	1	3	16
12	R_2lCOCM6Wmvp0gio	2	6.5	7	5	1	3	3	19
13	R_bO42S33cImaUeiM	1.5	7	7	5	3	3	3	17
14	R_3mVbCACxcnci0S0	1.5	7	6	3	1	2	3	7
15	R_5iiFkP25s9Lq89m	3	5.5	3	3	3	2	2	8
16	R_20s3J9eWQPHlkJ6	1	7.5	4	5	3	1	1	10
17	R_57JoCcWw7eqZI68	1	7.5	4	4	0	1	1	6
18	R_6J1YEgoo47Owwh6	1	7.5	5	2	2	1	2	10
19	R_298rSvTVBu0zNWI	1	7.5	7	4	3	3	2	8
20	R_1CflX3whac6Oeig	3	5.5	6	5	2	3	2	13
21	R_5ax87b7M88v06wI	1.5	7	6	5	3	3	3	15
22	R_efYOjGyGYllkenq	3	5.5	7	5	2	3	2	13
23	R_ePXrtbu84QOy8u0	3	5.5	7	5	2	3	2	12
24	R_cILgFHlM1gHqzdy	1	7.5	6	3	3	3	3	13
25	R_cAzmkCXiRVTxzww	1	7.5	5	3	2	1	3	12
26	R_9NAHtmvN4dUzxWs	3	5.5	1	4	2	1	0	10
27	R_b1Qr3rZZ6epbhly	1	7.5	4	2	2	2	1	13
28	R_ddnObNZSG7TvE9e	2	6.5	5	3	3	2	3	12
29	R_a4zv6qFVmejHlPK	5	3.5	7	4	3	3	3	16
30	R_3wnMpEv87HWDegk	2	6.5	7	4	3	3	3	12
31	R_9XHCPKlL3narVkg	1	7.5	7	4	3	3	3	12
32	R_55bsovRZTOFZgRS	1	7.5	6	4	2	3	1	12
33	R_5upApIoYf6ZFnkU	2.5	6	7	4	2	3	3	12
34	R_9pBzDL7K7UhJEVu	1	7.5	5	3	1	0	3	8

Case #	Case ID	DV	Reverse Recoded DV	IV1	IV2	IV3	IV4	IV5	IV6
35	R_0T9tkybSNAA79NG	1	7.5	5	4	3	2	3	13
36	R_3HF8w0E6XiZ8QS0	3.5	5	5	4	2	3	3	12
37	R_afclGk4uw5LIozG	2	6.5	5	2	3	2	3	10
38	R_bdvqoj6pQShGvVG	1	7.5	4	3	2	3	2	10
39	R_3Pi2Foom61MyLsw	3	5.5	4	3	2	0	2	6
40	R_cZU6NtqAqcGjmHG	4	4.5	6	4	2	3	1	14
41	R_7TWHjtCzzj1MbGI	1	7.5	6	3	2	1	2	13
42	R_beEr4v7YoazeGQQ	6	2.5	7	5	3	2	3	14
43	R_37C68fjypFDDUrO	4	4.5	4	3	1	0	2	7
44	R_9AoeP0h213NW3Iw	2	6.5	5	3	2	2	3	17
45	R_afSI9kIEo3DF7gM	1	7.5	5	3	3	1	3	14
46	R_39vGjDj70XGUgQs	3	5.5	6	3	1	2	2	7
47	R_eKWM5DfEXRS6Nog	1	7.5	5	3	2	3	3	12
48	R_bJDW4SsVxsWA192	2.5	6	7	4	3	3	3	11
49	R_6rgVpSHChf8Kp4E	1	7.5	5	3	3	1	3	16
50	R_9zP7tGBNPK9fmkI	1	7.5	4	3	1	2	3	6
51	R_e5w17AXiQErpEmo	2	6.5	7	3	2	3	2	13
52	R_2rfhpVZRoKyIP2I	1	7.5	6	3	3	3	3	14
53	R_2lvkhmbo61ktzrC	3	5.5	4	3	2	1	2	6
54	R_eQJUcH6a9JSQfT6	1.5	7	5	3	3	2	2	7
55	R_eFAG9ryOTReJlgE	2	6.5	7	3	2	2	3	6
56	R_a96hrZLpZipU5LK	2	6.5	4	3	2	3	2	11
57	R_1HXkwyW5C052ZA8	4.5	4	7	4	2	1	3	11
58	R_3meUXfgewhAOUv2	1	7.5	5	4	2	1	2	6
59	R_1ZjdR66ohqoxx88	4	4.5	7	5	3	3	2	12
60	R_3w7y1UlDoSDceJ6	1	7.5	5	4	3	0	3	9
61	R_2rTHXEDiC3HUqDa	7.5	1	4	3	1	3	1	7
62	R_5Ac5v8gqw3keaP2	1	7.5	7	3	2	2	3	12
63	R_9zPb7yIfN1LdJWc	1	7.5	6	3	2	3	2	13
64	R_5dyPrp5rj94gAHq	1.5	7	5	3	3	3	2	11
65	R_9HyawxnGJDH24EQ	1	7.5	5	3	3	0	2	11
66	R_emw09JmPfRo9xre	2	6.5	7	3	1	3	3	11

www.ingramcontent.com/pod-product-compliance
Lightning Source LLC
Chambersburg PA
CBHW081117170526
45165CB00008B/2471